Dynamic Data-Driven Simulation

Real-Time Data for Dynamic
System Analysis and Prediction

Dynamic Data-Driven Simulation

Real-Time Data for Dynamic
System Analysis and Prediction

Xiaolin Hu

Georgia State University, USA

World Scientific

EW JERSEY · LONDON · SINGAPORE · BEIJING · SHANGHAI · HONG KONG · TAIPEI · CHENNAI · TOKYO

Published by

World Scientific Publishing Co. Pte. Ltd.
5 Toh Tuck Link, Singapore 596224
USA office: 27 Warren Street, Suite 401-402, Hackensack, NJ 07601
UK office: 57 Shelton Street, Covent Garden, London WC2H 9HE

Library of Congress Control Number: 2022053777

British Library Cataloguing-in-Publication Data
A catalogue record for this book is available from the British Library.

DYNAMIC DATA-DRIVEN SIMULATION
Real-Time Data for Dynamic System Analysis and Prediction

ISBN 978-981-126-717-8 (hardcover)
ISBN 978-981-126-718-5 (ebook for institutions)
ISBN 978-981-126-719-2 (ebook for individuals)

For any available supplementary material, please visit
https://www.worldscientific.com/worldscibooks/10.1142/13166#t=suppl

Desk Editors: Logeshwaran Arumugam/Steven Patt

Typeset by Stallion Press
Email: enquiries@stallionpress.com

To my parents: Zejun Hu and Qiuying Duan

Preface

This book systematically presents dynamic data-driven simulation (DDDS) as a new simulation paradigm that makes real-time data and simulation model work together to enable simulation-based prediction/analysis. The term "Dynamic Data Driven Simulation" was first introduced in my 2011 *SCS Modeling and Simulation Magazine* article. Since then, it has become clear that a comprehensive description of DDDS needs to answer two key questions: (1) What is DDDS? (2) How to assimilate real-time data into simulation models? These two questions drive the main themes of this book.

A central task of DDDS is data assimilation. While data assimilation has been studied in other science fields (e.g., meteorology, oceanography), it is a new topic for the modeling and simulation community. A significant effort of this book is thus to describe data assimilation in a way that connects with the broad audience in the modeling and simulation field. This book bridges the two study areas of data assimilation and modeling and simulation, which have been developed largely independently of each other.

This book is the result of more than 10 years of research and development. I thank the students of the Systems Integrated Modeling and Simulation (SIMS) lab for their contributions to the work related to this book. The writing of this book is also helped by several colleagues. I am grateful to Bernard Zeigler for his guidance and Hessam Sarjoughian and Ming Xin for their inputs during the book writing process. A special word of gratitude is due to James Nutaro, who has read and commented on the book manuscript.

I thank my wife, son, and daughter for their support and patience during the long wait for the completion of this book.

It is hoped that this book will provide a comprehensive presentation of the DDDS topic and that it will serve as a reference and textbook for students and researchers working on the development and application of DDDS.

Xiaolin Hu

About the Author

Xiaolin Hu is a Professor of Computer Science at Georgia State University and heads the Systems Integrated Modeling and Simulation (SIMS) Lab. He obtained his Ph.D. in Computer Engineering from the University of Arizona in 2004. His research interests include modeling and simulation theory and application, dynamic data driven simulation, and complex systems science. He has numerous publications in leading scientific journals and conferences, and organized many international conferences and symposiums in the field of modeling and simulation. He was a National Science Foundation (NSF) CAREER Award recipient.

Contents

List of Tables

List of Figures

Chapter 1

Introduction

1.1 Why This Book?

The increasing amount of data collected from dynamic systems in real time poses a fundamental question to the modeling and simulation community: How to integrate real-time data with simulation to enable real-time prediction and analysis for a dynamic system under study? Answering this question requires a systematic examination of the relationship between data and simulation, and more importantly, new ways of making (real-time) data and simulation work together.

Simulation has long been used to study dynamic systems. Nevertheless, traditional simulation practices mainly view simulation as an offline tool to support system analysis, design, or planning. In these simulations, the simulation runs are based on historical data or hypothetical scenarios — they are not informed by real-time data collected from dynamic systems in operation.

The need for integrating real-time data with simulation is driven by two important trends in recent years. First, advances in sensor and communication technologies have significantly increased the availability and quality of real-time data collected from dynamic systems. These data carry information about the real-time conditions of systems, which can be utilized to make simulations more accurate. Second, as sophisticated simulation models have been developed, there is growing interest in using these models to support real-time decision-making for dynamic systems in operation (as exemplified by the increasing interests in *digital twin* technologies in recent years). The operational usage of simulation models in a real-time context requires simulation models to have the capability of providing real-time prediction/analysis for a system under study. Systematic ways of incorporating real-time data into simulation models are essential for achieving this goal.

The purpose of this book is to systematically introduce dynamic data-driven simulation (DDDS) as a new simulation paradigm, where a simulation system runs in parallel with a real system and constantly assimilates real-time data from the system. Presenting this new simulation paradigm requires an in-depth discussion of the related concepts and the methods for assimilating data into simulation models. The book addresses two key questions related to this topic:

- What is DDDS and how is it related to other modeling and simulation concepts?
- How to systematically assimilate real-time data into simulation models?

The first question is crucial because the modeling and simulation field has no shortage of new concepts and frameworks, and thus, it is important to clearly define DDDS in order to show its value. The need for integrating real-time data with simulation is easy to understand. Nevertheless, several key questions arise when one digs deeper into this topic. What are the data used in DDDS? What does DDDS exactly mean and what activities does it include? How is DDDS related to the existing modeling and simulation concepts? A major effort of this book is to systematically examine the connections between data and simulation, clearly define the concepts and activities of DDDS, and describe their boundaries as well as relationships with other modeling and simulation concepts. A set of DDDS activities are defined, including *dynamic state estimation, online model calibration, external input forecasting,* and *real-time simulation-based prediction/analysis.* Their roles in DDDS and their relationships with other modeling and simulation activities are described.

The second question deals with data assimilation methods for enabling DDDS. Data assimilation is a mathematical discipline that aims to combine a dynamic model with observation data to estimate the states of a system as it evolves over time. While data assimilation has been studied in other science fields (e.g., meteorology, oceanography), it is a new topic for the modeling and simulation community. This book bridges the two study areas of data assimilation and modeling and simulation, which have been developed largely independently from each other. The new paradigm of DDDS brings about the need to integrate them together. Special attention is paid to data assimilation for discrete simulations, which are the main type of simulations considered in this book. Examples of discrete simulations include discrete event simulation, discrete time simulation, and agent-based simulation. The discrete nature of these simulations poses unique challenges for applying data assimilation to discrete simulation models. These challenges are discussed and solutions are developed.

Integrating real-time data with simulation holds the promise of significantly increasing the power of simulation — it enables simulation-based prediction/analysis and allows simulation to play a more prominent role in supporting real-time decision making. To fulfill this promise requires the modeling and simulation community to work together to develop new methods, frameworks, and applications related to this topic. It is hoped that this book will promote more research and development and inspire new ideas in this field.

1.2 Scope and Structure of the Book

This book systematically introduces DDDS as a new simulation paradigm for dynamic system analysis and prediction. While DDDS is a general paradigm applicable to all simulations, the main focus of this book is on discrete simulations that use discrete simulation models. A significant portion of the book is dedicated to introducing data assimilation as an enabling technique for DDDS. Data assimilation deals with statistical estimation and is rooted in probability theory. An important aspect of data assimilation is the theoretical underpinnings of the various data assimilation methods. The material in this book focuses more on the concepts and procedures of data assimilation as opposed to the mathematical derivations and proofs of the data assimilation methods.

This book has been written with professionals, graduate students, and advanced undergraduate students in mind. It attempts to present the material in a way that requires a minimum of background knowledge. The book is suitable for both researchers and practitioners working in the related fields. Explanations of new concepts are accompanied by examples from diverse application areas.

A tutorial example is developed to provide a step-by-step guidance for setting up data assimilation and carrying out simulation-based prediction/analysis. A wildfire spread simulation application is also included to demonstrate how data assimilation and DDDS may be carried out in a more complex scenario.

Each chapter ends with a section describing the bibliography sources used by the chapter.

The book is organized into three parts, which are summarized as follows:

- **Part I: Foundation:** This part builds a foundation for introducing DDDS. It includes Chapters 2–4:

  - Chapter 2 describes the fundamental concepts of modeling and simulation, the relationship between data and simulation, and a comparison between data modeling and simulation modeling.
  - Chapter 3 introduces a taxonomy of simulation models and, based on that, describes the various simulation models and associated simulation algorithms.
  - Chapter 4 presents the basic probability concepts that are helpful for understanding the data assimilation methods.

- **Part II: Dynamic Data-Driven Simulation:** This is the main part of the book. It includes Chapters 5–7:

  - Chapter 5 systematically defines the new paradigm of DDDS and the DDDS activities.
  - Chapter 6 describes the data assimilation approach and the major data assimilation methods that include Kalman filter and particle filters. Special attention is paid to the treatments that are needed for working with discrete simulation models.
  - Chapter 7 presents a framework of DDDS for discrete simulations and provides a tutorial example demonstrating how particle-filter-based data assimilation can be carried out and how the different DDDS activities work together.

- **Part III: Application and Look Ahead:** This part builds on the previous chapters and describes a more advanced application and discusses some open topics for future research and development. It includes Chapters 8 and 9:

o Chapter 8 describes DDDS for the wildfire spread simulation application. It demonstrates the need for more advanced data assimilation when dealing with complex applications.
o Chapter 9 discusses open research questions and potential new topics related to DDDS.

Several examples are used throughout this book to help explaining/illustrating the various concepts related to DDDS. These examples come from different application domains. They include:

- wildfire spread simulation,
- road traffic simulation,
- mobile agent simulation,
- manufacturing system simulation,
- infectious disease spread simulation.

Part 1

Foundation

Chapter 2

Dynamic System, Simulation, and Data

2.1 Dynamic System and Simulation

Dynamic systems are everywhere. They range from the growth and adaptation of biological cells, to the congestion propagation and dissipation of highway traffic, and to planet movements in the solar system and beyond. They also cover different application areas, such as engineering (e.g., mobile robots), natural science (e.g., predator–prey ecosystems), social science (e.g., neighborhood segregation), economics (e.g., stock market), and public health (e.g., infectious disease spread). A common feature of these systems is that their states change over time, either driven by the systems' internal working mechanisms or external influences from the environment. The dynamic changes in states give rise to dynamic behavior and thus the name "dynamic system." This compares to *static systems* whose states do not change over time. Consider a building as an example. Its structure and height do not change and thus can be studied as a static system. On the other hand, its occupancy dynamically changes due to people's movement within the building and is best to be studied as a dynamic system.

People study dynamic systems for various reasons. From a scientific point of view, studying a dynamic system allows one to know more about the system and its environment. This satisfies human curiosity and helps in gaining new knowledge. From an engineering point of view, studying a dynamic system allows one to understand its behavior and working mechanism. This supports the operation of the system or designing new systems that work better. It also makes it possible to predict or analyze a system's future behavior to provide decision support.

A fundamental approach for studying dynamic systems is through experiments. An *experiment* is the process of extracting information from a system by exercising its inputs (Fritzson, 2004). An experiment can be carried out on the real system itself. For example, to evaluate how a new traffic control strategy (e.g., increasing the speed limit) works on highways, one can implement the traffic control on a real highway and observe its impact on traffic flows. Using a real system to do experiments, however, may face several practical challenges. These include the following: (1) the experiment might be too costly (e.g., replacing all the speed limit signs to do experiments can be costly); (2) the experiment might be too dangerous (e.g., increasing the speed limit can lead to severe car accidents); and (3) the experiment might be infeasible due to the fact that the real system does not exist or the experiment environment is not available (e.g., the specific highway under study is still being designed and does not exist).

A *simulation* is the imitation of the operation of a dynamic system through a model. This implies that a simulation allows an experiment to be carried out without using the real system itself; instead, it uses a model. The term *model* refers to an entity that is not the real system but resembles or conveys information about the behavior of the real system. A simulation exercise can take different forms depending on what model it uses. For example, a fire drill simulation would take a physical form as it uses physical objects and scenarios created in the physical world. This book focuses on *computer simulation* that uses computational models, which are models represented in an abstract way using mathematical forms or computer programs. Throughout the remainder of this book, the term "simulation" means computer simulation.

A simulation model is considered a type of mathematical model. Similar to other mathematical models, simulation models use abstract representations and are formally defined. Nevertheless, they have several unique characteristics that differentiate them from other mathematical models, as described in the following:

First, simulation models are *dynamic models* whose state values are time-dependent, i.e., changing over time. Time is an essential element of a simulation model. This compares to *static models* that are defined without involving time. An example of a static model is the equation describing Ohm's law: $I = V/R$, which specifies the static relationship between a conductor's current (I), voltage (V), and resistor (R).

Second, simulation models rely on computer simulation to generate results. A computer simulation starts from an initial state and iteratively computes the next state according to a time advance mechanism. It is generally considered a different approach from the *analytical approach* that derives solutions from equations. Consider the example of calculating the travel distance (d) of a vehicle that has an initial velocity (v_0) and a constant acceleration (a). Using the analytical approach, one can derive $d = v_0 t + at^2/2$, based on which d at any t can be directly calculated. A computer simulation, however, computes the results in an iterative way starting from an initial state. The advantage of step-wise computation of simulation is that it can effectively handle large-scale models that have no analytical solution. This compares to the analytical approach that often uses models based on simplified assumptions in order to make a problem tractable.

Third, many simulation models are expressed in forms that are different from conventional mathematical equations. Often, these models are defined by relations and rules that are embodied in computer programs. Consider Conway's *Game of Life* simulation (Gardner, 1970) as an example. The simulation model is described by a set of rules within a 2D cellular automata structure, as shown in Figure 2.1. The model needs to be simulated in a step-wise fashion in order to generate simulation results.

Simulation can be used in different ways for studying dynamic systems. Several common use cases of simulation are described as follows:

- **Theory and hypothesis test:** In this use case, simulation serves as an alternative to physical-world experiment to test some theory or hypothesis of interest. Simulation models are developed to realize the theory or hypothesis, and the developed models are

---

Conway's Game of Life Model:

The Game of Life is framed within a two-dimensional cell space structure. Each cell is in one of two possible states, *live* or *dead*. Every cell interacts with its eight neighbors, which are the cells that are directly horizontally, vertically, or diagonally adjacent. At each step in time, the following transitions occur:

- Any live cell with fewer than two live neighbors dies, as if by loneliness.
- Any live cell with more than three live neighbors dies, as if by overcrowding.
- Any live cell with two or three live neighbors lives, unchanged, to the next generation.
- Any dead cell with exactly three live neighbors comes to life.

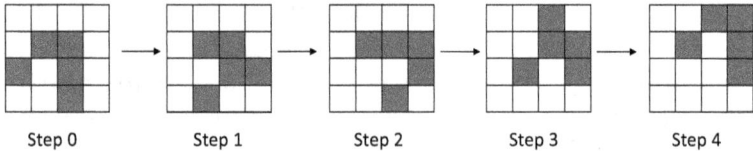

Step 0   Step 1   Step 2   Step 3   Step 4

Figure 2.1. Conway's Game of Life simulation: (top) description of the Game of Life model; (bottom) illustration of a step-wise simulation execution, where gray cells are *live* and white cells are *dead*. The illustration corresponds to five steps of simulation for the moving glider pattern.

simulated to check if the theory or hypothesis leads to expected results. Many simulations in the social science and natural science fields belong to this category. An important application of this use case is to study *emergent behavior*, which is system-level collective behavior formed from the self-organization of multiple simple entities (agents). Examples include the Game of Life simulation shown in Figure 2.1 and Reynolds' Boids model simulation (Reynolds, 1987).

- **System design:** In this use case, simulation is used to help in designing a new system that does not exist yet. The goal is to evaluate and test the different design choices using a digital version of the system. For example, before constructing a new highway interchange, simulation may be used to analyze the different interchange designs (e.g., standard diamond interchange, diverging diamond interchange) to help in choosing the one that works the best. Because the real system does not exist, a simulation model in this use case needs to be developed based on a design of the system. Often, multiple models need to be developed, which go through multiple iterations of development as the design process advances with more details.

- **System analysis and understanding:** In this use case, simulation is used to analyze an existing system to better understand its behavior and/or complexity. For example, simulation is commonly used to analyze road traffic of metropolitan areas. Another example is to use simulation to study the dynamic behavior of wildfire spread. A simulation model in this use case needs to be developed based on the real system under study. Depending on the analysis goals, different models may be developed to model different aspects of a system or to capture different levels of details of the system. Nevertheless, they should all reflect how the real system works.

- **System prediction:** In this use case, simulation is used to predict a real system's future behavior. A well-known example of this use case is weather forecast, which uses numerical weather models (a type of simulation model) to forecast future weather. We differentiate system prediction from system analysis because the former focuses on predicting the actual happenings of a real system and is typically run in a real-time context, whereas the latter is not limited by that. Despite their difference, simulation-based prediction and analysis often work together. To achieve accurate predictions, a simulation model needs to be developed at an abstraction level that adequately captures the dynamics related to the behavior under prediction.

The above use cases model and simulate a system under study itself. Another important usage of simulation is to provide "virtual environments" for other systems or users. These use cases model the environments within which other systems/users are tested or situated, as described in the following.

- **Simulation-based testing:** In this use case, simulation is used to provide a virtual environment to test a system under development. It is commonly used by engineers when developing embedded or autonomous systems, such as autonomous vehicles or robots. One of the main challenges of developing these systems is testing their perception, control algorithms, and actuation in a closed-loop fashion. Simulation-based testing makes it possible to carry out closed-loop testing without using a real environment. Simulation-based testing can take different forms, including software-in-the-loop or hardware-in-the-loop simulations.

- **Simulation-based training and education:** In this use case, simulation is used to provide realistic and immersive virtual environments that mirror real-life operations and scenarios to support learning and training. An example is a flight simulator that uses computer-simulated flight scenarios to train student pilots. Simulation-based training has the advantage that it is cost-effective. It also provides a safe and practical way for trainees to learn to handle adverse conditions that are too dangerous or impractical to reproduce in the real world.
- **Computer game and virtual/augmented reality:** Simulation is one of the enabling technologies for computer games and virtual/augmented reality. A unique feature of simulations in computer games and virtual/augmented reality is that they need to support interactions from players. Furthermore, as computer games become more and more network-based, it is important for the simulations to effectively handle simultaneous interactions from players across the network (e.g., the Internet).

The main focus of this book is on system prediction and analysis in a real-time context.

2.2   Framework for Modeling and Simulation

The diverse practices and applications of simulation call for a formal treatment of the modeling and simulation concepts. This section presents the modeling and simulation framework developed in Zeigler's *Theory of Modeling and Simulation* (Zeigler, 1976; Zeigler *et al.*, 2000). The framework defines the entities and their relationships that are central to computer modeling and simulation. It provides a foundation for understanding the other concepts to be developed in this book.

Figure 2.2 illustrates the modeling and simulation framework. The basic entities of the framework are *source system, model, simulator,* and *experimental frame*. The basic relationships among the entities are the *modeling relation* and *simulation relation*. More details for these entities and relationships are described as follows:

- **Source system:** The source system is the real or proposed system that we are interested in modeling and simulating. For a system that already exists in the real world, the source system is

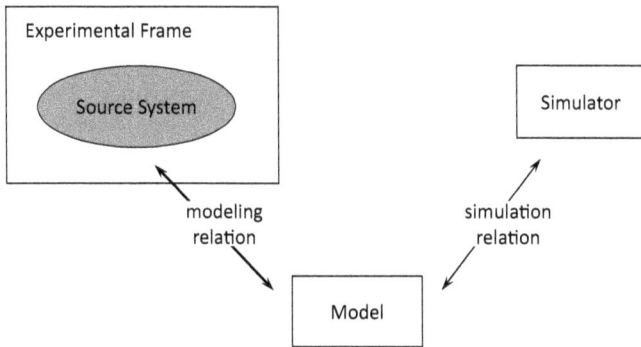

Figure 2.2. Basic entities and relationships in modeling and simulation. Reproduced from Zeigler *et al.* (2000).

also called the *real system*. The source system is viewed as a source of observational data or, more specifically, behavior. Knowing the source system is the starting point for developing a simulation model.

- **Model:** The model is the simulation model, which can be defined as a structure (e.g., a set of instructions, rules, equations, and/or constraints) for generating behavior resembling that of the source system. We distinguish the *structure* of a simulation model from its *behavior*. The structure refers to the elements of the model that make the model behave as it does; the behavior is the set of all possible data that can be generated from the model's structure.

- **Simulator:** The simulator is the computation system (i.e., a simulation algorithm) capable of executing a model to generate its behavior. Separating a model from its simulator brings several advantages, including supporting a modular design for simulation software, easier verification because a model and its simulator can be verified individually, and opening the way for portability and interoperability as the same model may be executed by different simulators.

- **Experimental frame:** An experimental frame specifies the conditions under which the source system is observed or experimented with. It is the operational formulation of the objectives that motivate a modeling and simulation project. An experimental frame can be viewed as a system that interacts with the system of interest to obtain data under specified conditions. In this view, a frame can be implemented to have the following three types of components: (1) *generator*, which generates input segments to the system; (2) *acceptor*, which monitors an experiment to see that the desired experimental conditions are met; and (3) *transducer*, which observes and analyzes the system's output segments.

The basic entities — source system, model, simulator, and experimental frame — are linked by two relationships:

- **Modeling relation:** The modeling relation links the source system and its model. It defines how well the model represents the source system. A model is considered *valid* if the data generated by the model agrees with the data observed from the source system in an experimental frame of interest.
- **Simulation relation:** The simulation relation links the model and the simulator. It represents how faithfully the simulator is able to carry out the instructions of the model. *A simulator correctly simulates a model* if it guarantees to faithfully generate the model's output trajectory given its initial state and input trajectory.

Several other concepts that are commonly used in the modeling and simulation literature are described as follows:

- **Modeling:** Modeling is the process of developing a model of a system. To differentiate from developing other types of models, the term *simulation modeling* is often used to refer to the process of developing a simulation model (the qualifier "simulation" may be omitted when the context is clear).
- **Simulation:** Simulation is goal-directed experimentation using a simulation model. It is a general concept referring to the process of using a simulator to execute a model to generate its behavior.
- **Simulation run:** A simulation run is a specific execution of a model by a simulator. During a simulation run, the sequence of external inputs received over time is called the *input trajectory*; the sequence of outputs generated by the model is called the *output trajectory*; and the sequence of states evolving over time is called the *state trajectory*. A simulation project typically needs to have many simulation runs. A simulation run may be simply called a simulation for brevity.

Similar to the software development lifecycle, the many modeling and simulation activities involved in a simulation project can be organized according to a *modeling and simulation lifecycle* (Balci, 2012). A modeling and simulation lifecycle defines the modeling and simulation activities from a process point of view. It typically includes the following stages that proceed in order: *problem formulation, conceptual modeling, simulation model implementation, verification and validation*, and *simulation-based experimentation*.

A modeling and simulation lifecycle may be divided into two broad phases: a *modeling phase* and a *simulation phase*. The modeling phase focuses on developing high-quality simulation models. Activities during the modeling phase are called *modeling activities*, which include all the activities related to conceptual modeling, simulation model implementation, and verification and validation. The simulation phase focuses on using simulation models to do experiments. Activities during the simulation phase are called *simulation activities*, which include all the activities related to simulation-based experimentation, such as experiment design, simulation execution, and result analysis.

2.3   Offline Simulation and Online Simulation

After a simulation model is developed for a source system, depending on whether or not a simulation is set up to run *concurrently* with the source system, we differentiate two types of simulation: *offline simulation* and *online simulation*.

Offline simulation is simulation that is independent of the real process of a source system. The independence means that an offline simulation does not use real-time data from the source system. This includes situations such as when the source system does not exist (e.g., it is still being designed), or the simulation experiment needs to simulate a scenario that is not connected to the real-time happenings of the source system. Offline simulation is also called *stand-alone simulation*. It is the traditional way of simulation, where simulation is used as a replacement for a real system for doing experiment. Offline simulation can be used to serve the following roles:

- **Supporting pure experimentation:** Offline simulation is commonly used to support experimentation for theory and hypothesis testing, system design, and system analysis. The experimentation typically includes many simulation runs corresponding to the different conditions or operation scenarios of a system. These experiments do not need to be connected to the real process of the system under study.

- **Supporting long-term planning or decision-making:** Long-term planning or decision-making needs to take into consideration the various scenarios that may happen. Offline simulation allows one to simulate the potential scenarios and analyze their results. For example, before each year's wildfire season, a fire manager may run simulations for various weather scenarios to generate a fire risk map, which is then used to support planning of firefighting resources before any wildfires actually happen.

- **Simulating historical events:** Simulations are used to reproduce or reconstruct historical events (e.g., a historical wildfire) to help understand what happened in the events. These simulations are unique in the sense that they are not experimentation based on hypothetical scenarios. Instead, they need to be set up using actual data available for the events. Historical event simulations belong to offline simulation because they do not run concurrently with the real events and they use historical data instead of real-time data.

Online simulation is simulation that runs concurrently with a source system and uses real-time data from the system (Ören, 2001). The term "online" implies that the source system is a real system in operation and that a simulation runs in parallel with the real process of the system. Often, there is also a real-time requirement for obtaining the simulation results, for example, in order to support real-time decision-making. Compared to offline simulation, online simulation is a relatively new way of running simulation. The dynamic data-driven simulation presented in this book belongs to online simulation. Online simulation can be used to serve the following roles:

- **Supporting real-time operation of a real system:** Online simulation allows system operators to simulate the dynamics of a real system to discover potential issues (e.g., bottlenecks in a manufacturing system) ahead of time, or to simulate the impact of an operational decision before applying it to the real system. These simulations provide useful information for the real-time operation of the system.

- **Supporting online performance monitoring and fault diagnosis:** By running a simulation model concurrently with a real system and comparing their behaviors, online simulation can support online performance monitoring and fault diagnosis (Ören, 2001). For example, a behavior discrepancy between the real system and the model may indicate a fault or malfunction in the real system.

- **Enabling real-time prediction:** By aligning a simulation with the real-time state of a real system (e.g., a spreading wildfire), online simulation can enable real-time prediction of the system's future behavior, e.g., to predict the wildfire spread. Such predictions can then support real-time decision-making for the systems under study.

It is important to differentiate online simulation from several other concepts. First, the term "online simulation" is sometimes used to refer to *live simulation*, which is an interactive simulation where real-world entities (e.g., people) may play the roles of some of the model components. For example, through live simulation a combat pilot can use a helmet to perceive virtual rival aircraft in an in-flight training session. The live simulation focuses on simulating a virtual environment for real-world entities. It is different from the online simulation presented in this book. Second, online simulation is different from *real-time simulation*, which refers to simulation where the simulation time is synchronized with the "wall-clock" time (see Chapter 3 for more details). Online simulation uses real-time data from a real system. However, it typically runs as fast as it can in order to meet the real-time requirement. Third, one should not confuse online simulation with simulations that happen through the Internet. Those simulations are often called *web-based simulation* or *cloud-based simulation*.

From a system development lifecycle point of view, offline simulation is often used during the design phase of a system to evaluate different design alternatives before a real system is developed, whereas online simulation is used during the operation phase of a system after the system has been developed and working in the real field.

2.4   The Rise of Data

Data are units of information representing facts of a system or a model. There has been a significant increase in both the quantity and quality of data in recent years due to technological advances on multiple fronts. On the sensor technology front, we are entering an era where sensors are ubiquitously deployed. A new generation of sensors provides more accurate measurements. They also have a longer battery life, a smaller size, and lower cost. This means more sensors can be deployed to collect more accurate data from dynamic systems. On the communication technology front, wireless communication makes it possible to gather data from places that are difficult to reach (e.g., in remote forests). The fast and reliable transfer also means data can be collected in real time and in high volume. On the storage technology front, multiple levels of data storage allow data to be stored locally in portable devices and in the cloud. The overhead of data storage and management is significantly reduced. Together, these technological advances have significantly improved the availability and reliability of data collected from dynamic systems. New technologies, such as Internet of Things (IoT) and 5G/6G wireless, are expected to continue this trend in the future.

Data collection methods vary by applications, ranging from manual procedures to technology-based methods. Collecting data through surveys or questionnaires, face-to-face interviews, and direct observation are examples of manual data collection. Technology-based data collection uses technology to support automated data acquisition. Examples include collecting data from physical systems using sensors and scaping data from websites using scraper bots (tools or programs for extracting data from web pages). This book focuses on data collection using sensors. A *sensor* is a device that converts a physical measure into a signal that is read by an observer or an instrument (Chen *et al.*, 2012). Sensors are commonly used to collect data from industrial machinery. They are also widely used to monitor systems' spatiotemporal behaviors, such as traffic monitoring on highways, air quality monitoring in metropolitan areas, wildfire detection and monitoring, and animal tracking in forests.

A wide variety of data have been collected from dynamic systems. In the following, we use two metrics to categorize the data for studying dynamic systems. First, depending on whether data change over time, they can be categorized into *static data* or *dynamic data*:

- **Static data do not change over time.** These data character-ize some static aspects of a dynamic system. Using the wildfire application as an example, data describing the terrain of a wildfire area are an example of static data because the terrain does not change during a wildfire event. Static data typically need not be referenced with the timestamp of when they were collected.
- **Dynamic data change over time.** These data are dynami-cally updated as new information becomes available. Examples of dynamic data in the wildfire example include the weather data that vary over time and the fire front location data that change as a wildfire spreads. Dynamic data need to be referenced together with the timestamp describing when the data were collected.

Dynamic data may be updated irregularly or, more likely, reg-ularly by the sensors that collect them. The regular updates of dynamic data may be characterized by an *update period* that defines the time interval between two consecutive updates or an *update fre-quency* (or *sampling rate*) that is the inverse of the update period. For example, a GPS sensor with a 2 Hz sampling rate means there are two GPS readings every second.

Data related to dynamic systems can also be categorized based on when they are collected and used. Based on this metric, the two cat-egories of data that play important roles in modeling and simulation are *real-time data* and *historical data*:

- **Real-time data are collected and used in real time.** They reflect the real-time situation of the system from which the data are collected. Real-time data are useful because dynamic systems evolve over time, and thus, any real-time analysis would need to take real-time data into consideration. Note that the term "real-time" typically does not mean "absolute real time," i.e., the data are collected at exactly the current moment. Instead, it means that the data are fresh data collected in the immediate past and have not been kept back from their eventual use after being collected.
- **Historical data are data collected in the past.** These data provide useful information for understanding a dynamic system to

be modeled, setting model parameters, and supporting calibration and validation of a developed model. It is important to know that historical data are in reference to when the data are used. Data that are collected at the time when they are used are real-time data. These same data would become historical data if they were used in a later time.

The two metrics described above can work together to categorize data with more granularity. Table 2.1 shows the data categorization based on the four combinations of the two metrics. For each sub-category, we provide examples based on the wildfire application context.

Each data sub-category described in Table 2.1 plays a unique role in modeling and simulation. For example, real-time dynamic data are needed for correctly initializing simulation runs in an online simulation; historical dynamic data are commonly used in simulation calibration and validation; and real-time and historical static data help characterizing a dynamic system for setting model parameters.

Table 2.1.   Data categorization and examples.

	Dynamic	**Static**
Real-time	**Real-time dynamic data**: data collected in real time that is related to a dynamic aspect of a system. *Example*: real-time wind speed data and wind direction data, real-time sensor data about fire front locations.	**Real-time static data**: newly arrived data that is related to a static aspect of a system. *Example*: newly arrived GIS data describing the terrain of a wildfire area, newly arrived fuel data describing the vegetation types of the wildfire area.
Historical	**Historical dynamic data**: data collected in the past that is related to a dynamic aspect of a system or an event. *Example*: fire front location data of a historical wildfire, wind data of a historical wildfire.	**Historical static data**: data collected in the past that is related to a static aspect of a system. *Example*: terrain data and fuel type data collected in the past, fire ignition location data for all the historical fires in the past 20 years.

Since real-time static data are static data that do not change over time, it is not important to emphasize the "real-time" property for the real-time static data. Due to this reason, the term "real-time data" generally only refers to real-time dynamic data.

2.5   Connecting Data to Simulation

Data play essential roles in almost every aspect of computer modeling and simulation. During the early stages of model development, data are used to formulate the design of a model and the environment in which the model resides. After a model becomes available, data are used to calibrate model parameters and to validate the model. Data are also used in a simulation run to set the initial state for a model, provide inputs to the model, and to record the behavior (i.e., output) of the model. This section discusses the connections between data and simulation. The discussion focuses on the simulation phase assuming that a simulation model has been developed.

The connections between data and simulation can be described based on the structure depicted in Figure 2.3. We identify four elements of a simulation model for which data are needed or produced during a simulation run, and show the corresponding data that are linked to them. The four simulation model elements are *initial state*, *model parameters*, *external input*, and *simulation output*. The data that are linked to these four elements as broadly named as the *system*

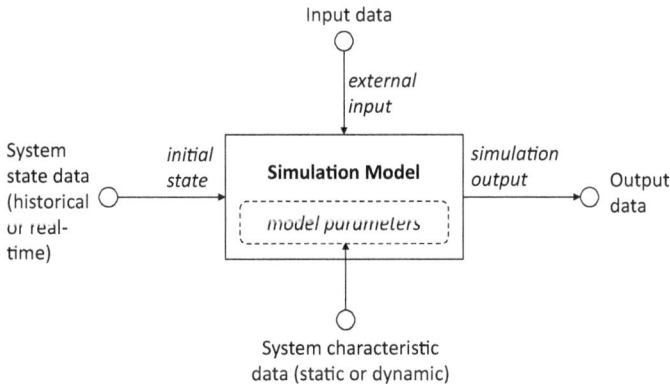

Figure 2.3.   Connecting data to simulation.

state data (data reflecting the system state), *system characteristic data* (data defining the system characteristics), *input data* (data defining the external inputs), and *output data* (data recording the simulation results), respectively.

Consider the example of a freeway traffic simulation where vehicles are modeled by agent models and the freeway is modeled as a road network. The initial state of a simulation would be set based on the data describing the initial traffic condition, e.g., the initial locations of vehicles. These data are system state data because they pertain to the state of the freeway traffic system. The parameters of the simulation model are set based on the data defining the traffic system's characteristics, such as the characteristics of the freeway road (e.g., length, number of lanes, speed limits) and the characteristics of the vehicles (e.g., types of vehicles and their preferred safe distances in car following). The external inputs are set based on the data defining the inputs to the freeway traffic system, such as inflow traffic on entrance ramps and control inputs to the ramp meters. The simulation outputs, such as the spatiotemporal traffic state and the output traffic on exit tramps, are recorded and stored as the output data.

In the following, we elaborate on the four model elements and their corresponding data.

2.5.1 *Initial state*

Initial state is the starting state of a simulation run, from which the simulation model evolves into new states and generates outputs over time. Different initial states would lead to different state trajectories and output trajectories. Thus, providing an appropriate initial state is essential for setting up a simulation run.

To set the initial state in a meaningful way, information about the source system's state, referred to as *system state data*, is needed. The system state data are broadly defined as any data that reflect the state of a system. Examples include observation data that directly capture a system's state (e.g., red or green state of a traffic light) or indirectly reflect a system's state (e.g., temperature sensor data that indirectly reflect the location of a wildfire) and data that describe general knowledge about how a system works (e.g., heavy traffic in the morning and light traffic during the night).

It is useful to differentiate historical system state data from real-time system state data. The former is collected from historical events. These data provide useful information for setting initial states in offline simulations. The latter is collected in real time from a system in operation. These data are needed for setting initial states in online simulations.

In principle, the initial state of a simulation should match the condition of the source system that is being simulated. This means the initial state needs to be set based on the system state data collected from the source system. For example, when simulating a historical wildfire, the initial state needs to be set based on the ignition or fire front locations recorded from the actual historical fire. Nevertheless, an important use of simulation is to analyze dynamic systems under various conditions. For these simulations, initial states are often set based on representative state conditions extracted from historical data that may or may not come directly from the source system. They may also be set based on whatever special conditions that a modeler would like to study. For example, a freeway traffic simulation project may initialize simulations with states corresponding to light, medium, or heavy traffic conditions and check how they influence traffic behavior. The light/medium/heavy traffic conditions are representative traffic conditions extracted from general traffic patterns observed in the past. As another example, a traffic simulation in a what-if analysis may initialize the model to a special state (e.g., a specific road segment is fully jammed) and study the system dynamics from there.

Sometimes, an initial state is set through a warm-up process by running a simulation starting from a special beginning state for a predefined burn-in time period or until a desired steady state is reached. For example, a city traffic simulation may start from an "empty" state with no vehicles on the road, run the simulation until all road segments reach desired levels of traffic, and then use that as the initial state for further experiment. This approach of using a warm-up process to set the initial state avoids the need for manually defining the initial traffic for each road segment, which can be a challenging task. The burn-in time period for the warm-up process depends on specific applications.

The above discussions mainly pertain to offline simulation. For online simulation, starting a simulation from the right initial state is even more important because one of the key requirements for online simulation is to make the simulation match the real-time condition of a real system. This means the initial state of an online simulation needs to be set based on the real-time observation data collected from the real system. This brings about the need for estimating the real-time system state from observation data. Addressing this need is one of the main activities of dynamic data-driven simulation that will be described in Chapter 5.

2.5.2 *Model parameters*

Model parameters are constants or variables that configure the specific form of a simulation model for generating behavior. Model parameters are internal to the simulation model. They need to have specific values in order to make a simulation model useful. Different model parameters will lead to different simulation results; thus, a critical step of a simulation study is to assign appropriate values to the model parameters. A simulation model's parameters and its state are two distinct concepts: The former is part of the model structure for generating behavior; the latter describes the condition of the model and is dynamically updated by the model structure. Consider the example of simulating a moving vehicle on a freeway. The vehicle's type (e.g., car or truck) and preferred safe distance in car following are examples of model parameters because they pertain to the characteristics of the vehicle. The vehicle's location and speed are examples of state variables, which are dynamically updated based on the simulation model.

Model parameters are determined or estimated based on *system characteristic data*, which are any data carrying information about the characteristics (e.g., properties, distinguishing features or quality) of a system under study. The majority of system characteristic data are static data. These data may come from sources such as system specifications, surveys, or field observations. For example, a freeway traffic system's road length, number of lanes, and speed limits can be obtained from the specification of the freeway; preferred safe distances in car following may be obtained from surveys of drivers; and the percentage of cars or trucks using the freeway may be obtained through field observations. These data may be processed by modelers to set the parameter values during the modeling phase.

Alternatively, they may be stored as data files and loaded by a simulation program at the beginning of a simulation run. The latter case allows a simulation model to be customized based on the loaded data. For example, by loading different road specification data, a freeway traffic simulation can simulate traffic on different freeway roads.

System characteristic data can also include dynamic data. One type of dynamic data is data describing the dynamic characteristics of a system. For example, the percentage of cars or trucks traveling on a freeway may dynamically change during a day. Thus, the data related to this characteristic of the system would be dynamic data. Another type of dynamic data is data describing the behavior of a system. An example of behavior data for the freeway traffic system is the spatiotemporal speed data regularly collected from different locations of a freeway road. These behavior data provide information based on which a model's parameters can be calibrated (i.e., adjusting the model parameters to make the model output match the observed behavior data). More on this topic comes later.

It is worth noting that a model's parameters may also be defined without using data directly. Some approaches are as follows:

(1) **Setting parameters based on general knowledge or personal experience:** This is the empirical way of setting model parameters, which may be used in the early stages of model development. For example, it is known that trucks generally have slower acceleration/deceleration rates compared to cars. This knowledge can be used to assign values for the acceleration/deceleration parameters of the two types of vehicles. A note of caution for this approach is that one's knowledge/experience may be incomplete or biased, and thus, extra attention needs to be paid to the corresponding parameters during model validation.

(2) **Setting parameters based on expert opinions:** This approach is commonly used when a modeler has limited knowledge of a system and there is limited data that can be used. In this case, one can consult domain experts and leverage their knowledge to set model parameters.

(3) **Setting parameters based on existing models:** When there are existing models or theories related to a model under development, one may reuse or adapt the model parameters from existing

models/theories. For example, when developing a new freeway traffic simulation model, one may check the car-following models and lane-changing models in the literature and use the corresponding parameters from those models to set the parameters of the new model.

2.5.2.1   *Model calibration and validation*

Two activities of the modeling phase that are closely related to model parameters are *model calibration* and *model validation.*

Model calibration can be defined as the process of finding a unique set of model parameters that makes the behavior of the simulation model match (i.e., "fit") the corresponding portions of observed data. Model calibration starts from an initial set of parameter values and then calibrates/adjusts the values based on observed behavior data. It can be treated as an optimization problem that aims to find parameter values to minimize the difference between the produced model outputs and the observed behavior data. The search for parameter values can be done through a variety of methods, including parameter sweeps (varying parameters and their combinations sequentially and systematically), search algorithms such as hill climbing and genetic algorithm, and even *ad hoc* approaches, such as manual tuning.

The underlying assumption of model calibration is that there exists a "fixed" set of model parameters that can make the model output fit a potentially wide range of observed behaviors. Once the model parameters are calibrated, they are "fixed," and the resulting simulation model is assumed to capture the characteristics of a system and thus can be used to study the system's behavior in other scenarios. This assumption may not work well in all applications. Sometimes, a system's characteristics change due to a shift in its operating environment. For example, the preferred safe distances in car following can change due to weather conditions (e.g., drivers tend to stay further away from the front vehicle during a storm). Situations like this are best modeled when the corresponding parameters can dynamically change to reflect the changing characteristics of a system. This creates the need for calibrating model parameters in an online fashion based on real-time observation data about a system's behavior. More details will be discussed in Chapter 5.

Model calibration often works together with model validation. Model validation refers to the process of determining whether the model, within its domain of applicability, behaves with satisfactory accuracy consistent with the modeling and simulation objectives (Balci, 1997). Model calibration and validation are closely related. Both are activities of the modeling phase to improve the credibility of a model before it is used. Both entail direct usage of data by comparing model output with observed data sets. Typically, a model needs to be calibrated before the validation activity is carried out. They may also be done in an iterative way to improve a model.

On the other hand, model calibration and validation differ in important ways, as described in the following:

- **The two deal with different problems.** Calibration aims to improve a model by adjusting the model parameters. Validation aims to evaluate if the agreement between model output and observed behavior is satisfactory. The validation should be carried out without any attempt to modify the model parameters to improve a model's fit.
- **The two use different data sets.** To ensure meaningful model validation, a validation should not use the same data used by model calibration. When historical data exist, part of the data may be used to build the model (including model calibration), and the remaining data are used to determine whether the model behavior agrees with that being observed in the real system (i.e., model validation).

2.5.3 *External input*

A dynamic system does not exist in a vacuum — it is influenced by other systems or the environment that are external to the system. When simulating the dynamic system, these external influences are modeled as input trajectories and fed into the simulation model. For example, a wildfire spread simulation is fed with weather inputs indicating changes in weather condition in the wildfire area; a manufacturing system simulation is fed with inputs about newly arrived jobs that need to be processed. The external input of a simulation model differs from the initial state and model parameters in two important ways. First, input represents influences external to a dynamic system; the values of input are determined outside the system. This compares to the other two elements that describe the condition and characteristics of the dynamic system itself. Second, input manifests itself through an input trajectory, which is a sequence of external inputs over time.

The input of a model is linked to *input data*, which are data describing the external inputs of a source system. These data are dynamic data collected by sensors or other data acquisition methods. Based on existing input data, the input trajectory of a simulation run can be defined using the following two approaches: (1) using the actual input data and (2) input modeling:

- **Using the actual input data:** In this approach, the actual input data act as the input trajectory for a simulation run. This approach is typically used when simulating historical events or when carrying out analysis against past input scenarios. For example, when simulating a historical wildfire, the actual weather data for the fire event would be used as the weather input for the simulation.
- **Input modeling:** In this approach, a model of input is developed based on the available input data. The developed input model is then used to generate input trajectories for simulation runs. A common method of input modeling is to select and fit a probability distribution to the input data and then use the fitted probability distribution as the input model (Biller and Gunes, 2010). For example, an exponential distribution may be used to model the time to the next job arriving at a manufacturing system, and the rate parameter of the exponential distribution can be determined by fitting the distribution to historical job-arrival data.

When input data is not directly available, an input model may still be developed using information from other sources, such as knowledge about the input process itself or expert opinions. For example, simulating wildfire spread in extreme wind conditions needs input about extreme wind speeds. The wind speed range (e.g., 80–120 km/h) corresponding to extreme wind conditions for wildfire spread may be obtained from domain experts. This information can be used to develop an input model that generates input trajectories of extreme wind speeds for wildfire spread simulations.

2.5.4 *Simulation output*

A simulation generates output, which are time-indexed simulation results produced by the simulation model. The output of a simulation run includes both the output trajectory and state trajectory. The former is output that is externally observable — a model generates its output through an external interface or output ports. The latter is the internal state of the model. Both the output trajectory and state trajectory are dynamically computed as the result of a simulation run.

Output data are data that stores the output and state trajectories of a simulation run. Different from the previous input data that are used by a simulation, the output data are generated from a simulation. When collecting output data, one should make sure that the data recording procedures do not interfere with or alter the behavior of the simulation model.

Simulation output analysis is the analysis of output data to evaluate a system's performance or to make inferences about a system's behavior. In order to produce statistically sound analyses, output analysis typically uses data from many simulation runs. The different simulation runs may have different initial states, parameters values, and/or inputs trajectories. The results of the many simulation runs are statistically analyzed, for example, to compute the mean, standard deviations, and confidence intervals for some measurements. Related to output analysis is *simulation visualization*, which uses data visualization techniques to visually represent output data to make the simulation results easier to understand, interpret, and analyze. This is useful for presenting complex simulation results and communicating with other users, such as decision makers.

2.6 Combining Simulation Model and Real-Time Data

The real-time data collected from a system carries important information about the system. Considering a real system in operation, we now have two broad sources of information: *knowledge* about how the system works, which is embodied by a simulation model; and *measurements* of the system's real-time operation, which are captured by real-time data. The real-time measurement data are also called *observation data*, or simply *observations*, as they are related to the observability of the system.

Both simulation model and measurement data have limitations. A simulation model is an abstraction of a real system. The accuracy of a simulation depends on the fidelity of the simulation model as well as other factors, such as the data used for setting model parameters or inputs. Consider the wildfire spread simulation as an example. The simulation model captures the principles of fire spread; however, it has errors due to its mathematical representation of a real physical process. The terrain data, fuel data, and weather data used by the simulation model also have errors. For example, terrain and fuel data typically have limited spatial resolution (e.g., 30 m). This means the terrain and fuel within the spatial resolution are considered uniform, while in reality they vary spatially. Similarly, weather data are often obtained from a nearby weather station in a time-based manner. Before the next data arrives, the weather is assumed unchanged. This is different from reality, where the real weather constantly changes over time. Due to these errors, the results from a simulation model will inevitably be different from what is observed in a real fire spread.

Real-time measurement data have errors too. The measurements can be inaccurate or noisy due to sensor limitations. Furthermore, measurements are often discrete in space and time, meaning that they do not have complete information regarding how a system works. For example, drones equipped with thermal cameras may be used to collect real-time data about the spread of a wildfire. Nevertheless, the measurement data may be noisy (e.g., due to low resolution of the thermal camera) or partial (e.g., a thermal image at any time covers only a portion of a large wildfire).

Combining the two sources of information from a simulation model and real-time measurement data has the potential to make simulation a more powerful tool for studying dynamic systems. The real-time data make it possible to align a simulation more closely with a real system and to dynamically adjust the simulation model to generate more accurate results. Doing this in a systematic way leads to a new simulation paradigm called *dynamic data-driven simulation*, as illustrated in Figure 2.4. Note that since simulation is run in parallel with a real system and uses real-time data, dynamic data-driven simulation is a type of online simulation. Nevertheless, it goes beyond conventional online simulation by emphasizing continuous

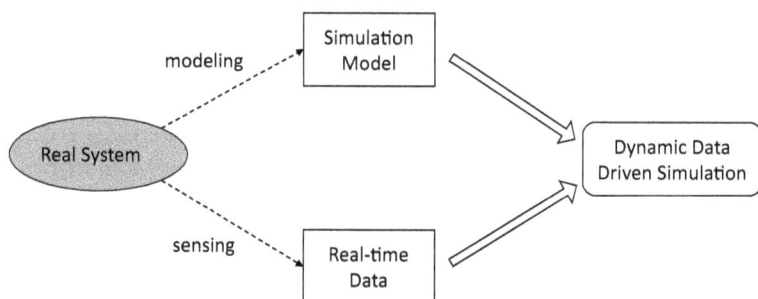

Figure 2.4. Combing simulation model and real-time data for dynamic data-driven simulation.

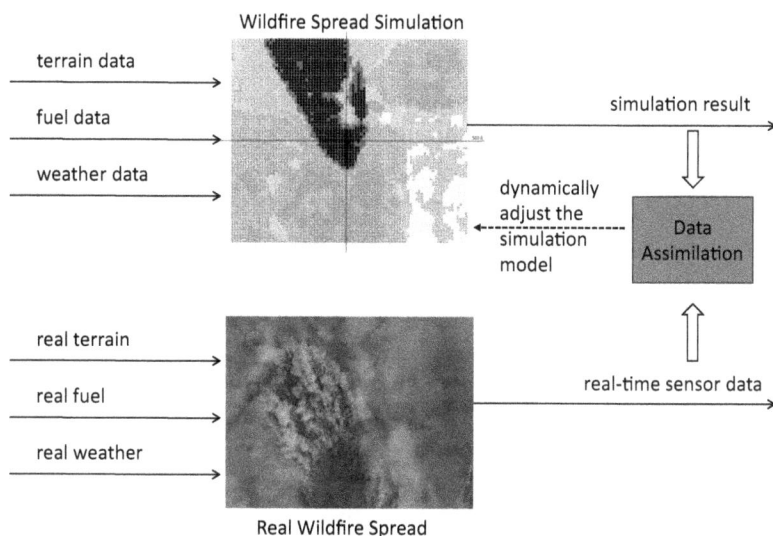

Figure 2.5. Dynamic data-driven simulation for wildfire spread prediction.

and systematic incorporation of real-time data. More details will be provided in Chapter 5.

Figure 2.5 illustrates the idea of dynamic data-driven simulation for the wildfire spread simulation application. In the figure, the top component represents the wildfire simulation model and the bottom component represents a real wildfire. As the wildfire spreads, streams of real-time data are collected from fire sensors, e.g., ground temperature sensors deployed at various locations of the fire area. These data reflect the dynamically changing state (e.g., fire front locations) of the wildfire. The real-time sensor data are assimilated by the data assimilation component. The results of data assimilation are used to dynamically adjust the simulation model to make the simulation more accurate. For example, based on the real-time sensor data of a wildfire, the data assimilation can estimate the real-time fire front locations and use that information to initialize simulation runs for more accurate simulation-based prediction/analysis. The sensor data also carry "feedback" information for a simulation to dynamically calibrate its model parameters to reduce the discrepancies between simulation results and observed fire spread behavior.

2.7   Data Modeling vs. Simulation Modeling

The dynamic data-driven simulation uses real-time data by assimilating them into a simulation model. An alternative approach of using data is to exploit data directly for patterns, trends, and correlations. This approach has gained much attention in recent years due to the big data trend that has resulted in large data sets that can be exploited. We refer to this approach as *data modeling* and differentiate it from *simulation modeling*. A comparison between the two approaches will help in understanding their differences and applicability. The material in this section is influenced by the work of Kim *et al.* (2017).

Data modeling, also called *modeling with data*, is a process of extracting and discovering meaningful patterns, trends, and correlations from large data sets. The product of data modeling is a *data model* that can be used to predict outcomes that may not necessarily be in the original data sets, e.g., outcomes in the future. Data modeling has achieved great success in applications such as image recognition and product recommendation. It is also increasingly used for predicting/analyzing behaviors of dynamic systems. Data modeling can work for dynamic systems because the data collected from these systems are the results of the systems' internal working mechanisms and principles. As long as the mechanisms and principles do not change, similar patterns, trends, and correlations can be expected in the future. In other words, the past can predict the future. For example, data modeling may discover traffic patterns from historical data and use them to predict future traffic.

Data modeling can be carried out using a range of methods, among which the most important one is *machine learning*. Machine learning is a type of artificial intelligence (AI) that has the ability to learn from past data to predict the future. Take machine learning based on artificial neural network (ANN) as an example. An ANN has a general structure that consists of nodes and connecting edges, whose parameters (e.g., weights) can be trained from historical data. When applied to a dynamic system, an ANN can be trained to learn the relationship between the system's inputs (or input trajectories) and its outputs based on historical data of the system's behavior. The trained ANN can then be used to predict the system's future output by feeding it with the corresponding inputs.

Data modeling uses past data to predict the future. This speaks to both its strength and weakness, as will be discussed in more detail later. To make data modeling work in a robust way, having large data sets is essential because large data sets are more likely to cover the patterns and correlations that can be used for predicting the future.

On the other hand, simulation modeling is a process of creating a simulation model that models the processes and mechanisms, i.e., the physics, governing how a dynamic system works. To make simulation modeling work, a modeler needs to have knowledge about how the system works. A simulation model is a "white box" model because it explicitly models the internal mechanisms of a system. In contrast, a data model is a "black box" model because it captures the input–output correlation from an external point of view.

Figure 2.6 shows an example to compare the two modeling approaches. The dynamic system under consideration is a round-about intersection connecting three two-way traffic roads (top image in the figure). A roundabout intersection is a circular intersection where drivers travel around a central island. There is no traffic signal; drivers yield to traffic at entry into the roundabout, then enter the intersection and exit at their desired street.

A particular interest in studying a roundabout intersection is to analyze how its inflow traffic (i.e., the number of arriving vehicles per time unit) influences its outflow traffic (i.e., the number of departing vehicles per time unit). To study this using a data modeling approach, an ANN model may be developed to have an input layer with three input nodes corresponding to the traffic of the three inflow roads, a hidden layer with four internal nodes, and an output

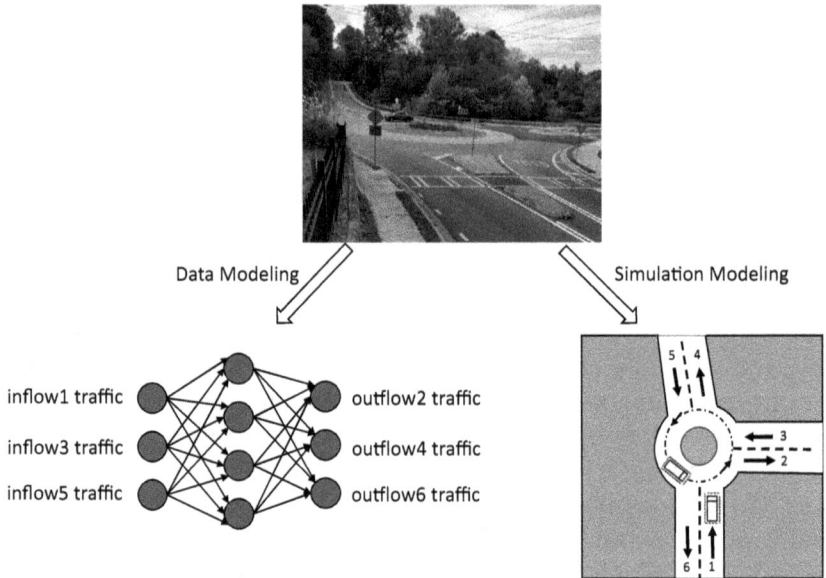

Figure 2.6.   Data modeling and simulation modeling for studying a roundabout intersection: (top) the roundabout intersection, (bottom left) a data model, and (bottom right): a simulation model. Each road (inflow or outflow) is assigned an identifier, as shown in the bottom-right diagram of the figure. The inflow roads are denoted by *inflow1*, *inflow3*, and *inflow5*, and the outflow roads are denoted by *outflow2*, *outflow4*, and *outflow6*.

layer with three output nodes corresponding to the traffic of the three outflow roads. This is illustrated in the bottom-left diagram in Figure 2.6. The model would be trained using historical traffic data reflecting the relationships between the inflow traffic and the outflow traffic. Note that the hidden layer and its four internal nodes are part of the hyperparameters of the ANN, which bear no resemblance to the structure of the intersection itself. In fact, an ANN could have multiple hidden layers, each of which has a different number of nodes. On the other hand, to study the intersection using a simulation modeling approach, a simulation model would be developed to explicitly model the intersection's layout and size, traffic rules, and vehicles' moving behaviors based on their arrival and departure roads. This is illustrated in the bottom-right diagram in Figure 2.6.

Table 2.2 summarizes the major differences between the two modeling approaches using three metrics that are expressed in

Table 2.2.   Comparing data modeling and simulation modeling.

	Data modeling	Simulation modeling
What is needed for developing a model?	Data (a model is trained through data)	System knowledge (a model is constructed based on system knowledge)
What does a model represent?	"Black-box" model representing correlation	"White-box" model representing causality
What can a model do?	Predictive analysis	Predictive analysis and prescriptive analysis (including what-if analysis)

three questions. Details about these differences are elaborated upon in the following.

2.7.1   *What is needed for developing a model? Data vs. system knowledge*

Data modeling relies on data. It can proceed as long as there are sufficient data. This can be advantageous when there are sufficient data but knowledge about how a system works is lacking. However, the reliance of data modeling on data also brings about several limitations. First, data modeling does not work well when data are not available or data are scarce. An example of unavailable data is when studying the impact of a new traffic rule that has never been implemented before; an example of scarce data is when studying the COVID-19 disease spread at the beginning of the COVID-19 pandemic. Second, data modeling is not effective in handling rare events (e.g., crashes of the stock market, stampedes in crowded environments). This is because the rarity of the events means that there would be a lack of quality data for training a data model for these events. On the contrary, a simulation model can generate reasonable results for rare events as long as there is a good understanding about the dynamics of the events. Third, the quality of training data is vital for data modeling. Unfortunately, data may be incomplete, biased, or altered (intentionally or unintentionally). Collecting reliable or trustable data is not always an easy task. This limits the applicability of data modeling.

Simulation modeling uses knowledge about how a system works. This makes it a viable approach even when data are not available. For example, one can develop a simulation model based on disease spread principles to study a new infectious disease even when related data are scarce. On the other hand, the need for system knowledge indicates a limitation of simulation modeling: It does not work well if there is insufficient knowledge. For example, developing a simulation model to simulate how our brain recognizes images remains an unachievable task due to the insufficient knowledge about how the brain works for image recognition. This compares to the data modeling approach that has gained success in image recognition thanks to the existence of a large amount of data. One way to address the issue of insufficient knowledge is to make hypotheses and then develop models based on the hypotheses. In this way, one should explicitly document the hypotheses and validate the developed models using data whenever possible.

2.7.2 *What does a model represent?*
Correlation vs. causality

A fundamental difference between data modeling and simulation modeling is that data modeling models correlation while simulation modeling models causality. Here, *correlation* means associational relationship between variables, and *causality* means cause–effect relationship between variables and processes, i.e., why and how things happen (Kim *et al.*, 2017). A data model represents the correlation between its input and output. Modeling correlations can be useful because they indicate relationships that can be exploited in practice for prediction. Consider the roundabout intersection in Figure 2.6, for example. One may observe a negative correlation between *inflow1 traffic* and *outflow2 traffic*: When there are many vehicles coming from the *inflow1* road, the traffic on the *outflow2* road decreases. This correlation, regardless of the mechanisms that cause it, may be used to predict the *outflow2 traffic* if one knows the traffic on the *inflow1* road.

On the other hand, a simulation model models the causality of actions or events that happen over time. The sequence of actions or events happening due to the causal relationships gives rise to the dynamic behavior of the simulation model. For the above-mentioned roundabout intersection example, the negative correlation between *inflow1 traffic* and *outflow2 traffic* could be due to the fact that the majority of vehicles coming from the *inflow1* road need to use the roundabout intersection to make U-turns. An increased *inflow1 traffic* thus would make the intersection congested and result in reduced *outflow2 traffic*. Through simulations, such a causality relationship between *inflow1 traffic* and *outflow2 traffic* can be revealed, and the negative correlation between the two can be explained.

The above discussion exposes a major limitation of data modeling: Interpretation of results from a data model may be difficult. As a "black box" model representing input–output correlations, a data model does not provide information to explain *why* such correlations exist. This compares to a simulation model that models causality, based on which the dynamics and correlations can be explained. For many applications, especially applications involving decision-making or policy-making, understanding the why is crucial in order to reach a level of knowledge that can be used with confidence (Mazzocchi, 2015). The lack of interpretability and understandability remains a challenging issue for data modeling. The focus on correlation instead of causality in data modeling also means that a data model cannot cope with anomalies and changing circumstances of a system. A data model works only in the same condition as defined by the data used for training the model.

2.7.3 *What can a model do? Predictive analysis vs. prescriptive analysis*

The two types of models can also be compared by looking at what they can do. Following the work of Kim *et al.* (2017), we align this comparison with the three levels of system analysis, namely, *descriptive analysis*, *predictive analysis*, and *prescriptive analysis* (Evans, 2016). The descriptive analysis explains what has happened in the past. The (historical) data sets as well as visualizations of the data sets support analysis at the descriptive level. The predictive analysis is a more advanced analysis that predicts what will happen in the future. Both data models and simulation models support analysis at the predictive level. A data model can support prediction by taking the input data in the immediate past to predict the output in the near future. A simulation model supports predictive analysis by running simulations to generate output in the future based on the dynamics modeled by the model. Real-time prediction can also be achieved through dynamic data-driven simulation that incorporates real-time data into simulation runs.

The prescriptive analysis takes the predictive analysis to the next level. It analyzes various actions and interventions that may be applied to a system and evaluates their outcomes and potential implications. A data model cannot be used for prescriptive analysis as it only models the input–output correlation. On the other hand, a simulation model supports prescriptive analysis as the various actions and interventions can be added to the simulation model and their impacts simulated and evaluated. An important type of analysis is the "what-if analysis," which inspects the behavior of a dynamic system under some hypothetical scenarios. By definition, what-if analysis is at the prescriptive level as it requires changing the operating conditions of a system to evaluate the consequences. Simulation models are commonly used to support what-if analysis for complex dynamic systems.

2.8 Sources

The modeling and simulation concepts introduced in this chapter are derived from various previous works in the modeling and simulation literature, which have been repackaged or elaborated on to serve the needs of this book. Among the various references, some play a more direct role in helping present the material in this chapter. In particular, the introduction of dynamic system and simulation in Section 2.1 is influenced by Chapter 1 of Fritzson (2004). The framework for modeling and simulation in Section 2.2 is based on the foundational work of Zeigler (1976) and Zeigler *et al.* (2000). The concepts of offline simulation (i.e., stand-alone simulation) and online simulation in Section 2.3 were adapted from the work of Ören (2001). The introduction of dynamic data-driven simulation in Section 2.6 is adapted from Hu (2011). Finally, the comparison between data modeling and simulation modeling in Section 2.7 is influenced by Kim *et al.* (2017).

Chapter 3

Simulation Models and Algorithms

This chapter provides an overview of the major types of simulation models and corresponding simulation algorithms. For each model and algorithm, we focus on its basic form while skipping the various variations or extensions that may exist.

3.1 A Taxonomy of Simulation Models

A wide range of simulation models exist, each of which has unique features that are suitable for specific applications. The many simulation models can be categorized using different categorization criteria. Figure 3.1 shows a taxonomy of simulation models according to the following three criteria: (1) *timing of change*; (2) *aggregation level*; and (3) *randomness*. Using the timing of change criterion, simulation models can be categorized as *continuous*, *discrete time*, or *discrete event* models. Using the aggregation level criterion, simulation models can be categorized as *micro-level* or *macro-level* models. Using the randomness criterion, simulation models can be categorized as *stochastic* or *deterministic* models. Each category also lists some example models. More details about these example models will be described in later sections.

The taxonomy tree in Figure 3.1 shows that a model can appear in multiple categories depending on what criteria are used. For example, a differential equation model is a continuous model according

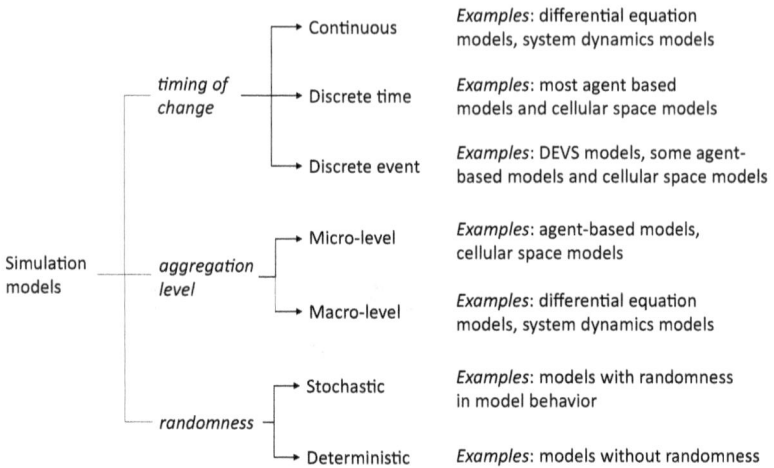

Figure 3.1.   A taxonomy of simulation models.

to the timing of change criterion; it is considered a macro-level model according to the aggregation level criterion; and it can be a stochastic model or a deterministic model depending on whether or not the model has randomness (i.e., noise factors). Similarly, an agent-based model is a micro-level model from the aggregation level point of view. From the timing of change point of view, an agent-based model may be a discrete time model or a discrete event model, depending on whether the model is implemented in a discrete time fashion or a discrete event fashion.

Models may also be categorized in other ways that are not shown in Figure 3.1. For example, one may use a "connection topology" criterion to further categorize agent-based models based on how agents are connected with each other. Agents can be connected through a Euclidean space (e.g., a 2D or 3D continuous space), or a discretized cell space (such as in the Game of Life simulation), or through social networks.

3.1.1   *Timing of change*

The timing of change criterion categorizes simulation models based on how they reflect state change over time. The three categories of models under this criterion are *continuous model*, *discrete time model*, and *discrete event model*. In continuous models, changes in state happen continuously over time (time is a real number). In discrete time models, changes in state happen at discrete time steps (time is an integer). In discrete event models, changes in state are driven by the occurrences of events, which can happen at any time over a continuous time base (time is a real number). The three categories of models are specified by three basic modeling formalisms (Zeigler *et al.*, 2000): Continuous models are specified by *differential equations*; discrete time models are specified by *difference equations*; and discrete event models can be specified by the *discrete event system specification* (DEVS) formalism, as will be described later.

The timing of change has a fundamental impact on the dynamics of a simulation model. The different mechanisms of timing of change lead to different forms of state trajectories. This is illustrated in Figure 3.2. Specifically, continuous models have a continuous state and time base; the state trajectory is a continuous curve over time, as illustrated in Figure 3.2(a). For discrete time models, the time base is discretized into fixed time steps, and the state is updated only at the discrete time steps. This is illustrated in Figure 3.2(b). For discrete event models, the time base is continuous and the state is updated at discrete time points associated with events. Events can happen at any time, and the time intervals between events are irregular. The state trajectory of a discrete event model thus consists of piecewise constant segments, as shown in Figure 3.2(c).

All three types of models are widely used. Continuous models are commonly used to model physical processes of natural or engineering systems, such as the atmospheric processes of a weather

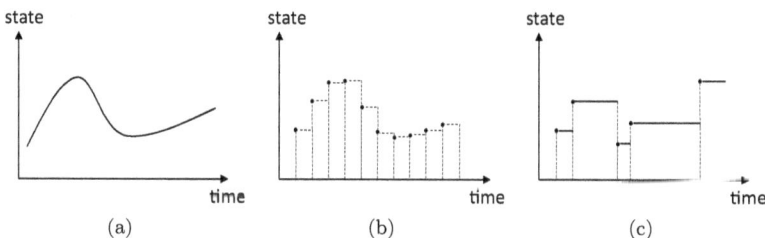

Figure 3.2. Time base and state trajectory of simulation models: (a) continuous model — time base is continuous, and state changes continuously over time; (b) discrete time model — time base is discrete, and state changes at regular time steps; (c) discrete event model — time base is continuous, and state changes at discrete time points that have irregular time intervals.

system and the trajectory of a flying missile. Examples of continuous models include differential equation models and system dynamics models. Discrete time models find applications in domains such as biological system simulation, social simulation, artificial society simulation, and robotics simulation. Most agent-based models and cellular space models are implemented as discrete time models. Discrete event models are commonly used to simulate applications such as queuing systems in service applications, manufactory production systems, inventory management, control systems, and computer networks, where the arrival and completion of messages, calls, jobs, or entities are discrete events in nature. Examples of discrete event models include DEVS models and agent-based and cellular space models that are implemented in a discrete event fashion.

Among all the categorization criteria, the "timing of change" criterion is a special one because it dictates different simulation algorithms for simulating the three categories of models (continuous, discrete time, and discrete event). Details of these simulation algorithms will be provided in later sections. To simulate models that are categorized under other criteria, one needs to check the "time of change" mechanisms of the models and then select the corresponding simulation algorithms to simulate the models. For example, when an agent-based model is implemented in a discrete time fashion (i.e., agents update their states at fixed time steps), it needs to be simulated using a discrete time simulation algorithm. When an agent-based model is implemented in a discrete event fashion (i.e., each agent updates its states based on event occurrences), it needs to be simulated using a discrete event simulation algorithm.

Simulation of continuous models is generally referred to as *continuous simulation*. To differentiate from continuous simulation, this book uses *discrete simulation* to refer to simulations of discrete time models or discrete event models. Discrete time models and discrete event models are also called *discrete simulation models*.

3.1.2 *Aggregation level*

The aggregation level criterion categorizes simulation models based on how they represent a system and its entities. Under this criterion, models are categorized as *macro-level models* or *micro-level models*, which are results of the *macroscopic modeling* approach and *microscopic modeling* approach, respectively. The macroscopic modeling approach models a system's aggregate or collective behavior as a whole without focusing on the events and interactions occurring at the individual level. On the other hand, the microscopic modeling approach models the behaviors of individual entities; it takes the viewpoint that a system is made up of individuals that have their own behaviors. The two modeling approaches differ in the perspective of modeling a system: The macroscopic modeling uses a top-down perspective to model a system, whereas the microscopic modeling uses a bottom-up perspective to model a system.

Consider freeway traffic simulation as an example. Macro-level models would model the aggregate behavior of the traffic flow. They may define the spatial density and average velocity as a function of the freeway location and time. These models generally assume homogeneous vehicles that are well mixed. An advantage of macro-level traffic models is that they can simulate some of the traffic phenomena without modeling all the details of a traffic system. In contrast, micro-level models would consider the behaviors and interactions of individual vehicles. They may include parameters describing each vehicle's location, speed, destination, and driving behavior (e.g., car-following and lane-changing behavior). These models can easily incorporate heterogeneity in vehicles' attributes. However, the increased detail comes at the cost of using large numbers of parameters as well as higher computation costs for simulating the models.

Macro-level models are often expressed as differential equations. System dynamics models are also macro-level models because they model stocks and flows of a system at the aggregation level. Micro-level models need to model each individual and its interaction with the environment. Agent-based models and cellular space models are classic examples of micro-level models. Discrete event models can be considered micro-level models because they model the events associated with sub-components of a system. Nevertheless, one needs to be aware that discrete event models are classified based on the "timing of change" criterion as opposed to the "aggregation level" criterion.

3.1.3 *Randomness*

The randomness criterion categorizes simulation models based on whether or not they involve randomness in model behavior. The two categories of models under this criterion are *deterministic models* and *stochastic models*. A deterministic model is a model in which the input and state uniquely determine the next state and output. A stochastic model involves randomness in its model behavior, meaning that the model's output is not uniquely determined by the input and state. Consider a manufacturing system as an example. In a deterministic model, one would, for instance, assume that new jobs arrive every 10 min and it takes 7 min to finish processing each job. Thus, given an initial state of the manufacturing system, the model's future state (e.g., number of jobs in the queue) and output (e.g., each job's finishing time) is uniquely defined. On the other hand, in a stochastic model, one would model the arrival time and processing time following some probability distributions. For example, the job processing time may follow a normal distribution $N(7, 4)$ that has a mean of 7 (min) and a variance of 4. The randomness built into the model means that the number of jobs that will arrive in the future as well as the time it takes to finish each job cannot be exactly predicted.

The two types of models have important implications for the simulation results. For deterministic models, different simulation runs with the same initial state and input trajectory would always generate exactly the same state trajectory and output trajectory. This is not the case for stochastic models. A stochastic model would generate different results for the different simulation runs even though they start from the same initial state and receive the same input trajectory. This is due to the randomness that is involved in the model behavior. Figure 3.3 illustrates this difference between the two types of models. Due to the different results from the different simulation runs, when simulating stochastic models, a large number of simulation runs are often used in order to draw meaningful conclusions from the results.

Stochastic models need to use random numbers in their model implementations. The common approach is to use *pseudorandom number generators* provided by the corresponding programming languages or simulation packages. A pseudorandom number generator is able to produce a sequence of numbers that imitates the ideal properties of a random number distribution. An advantage of a pseudorandom number generator is that the sequence of generated random

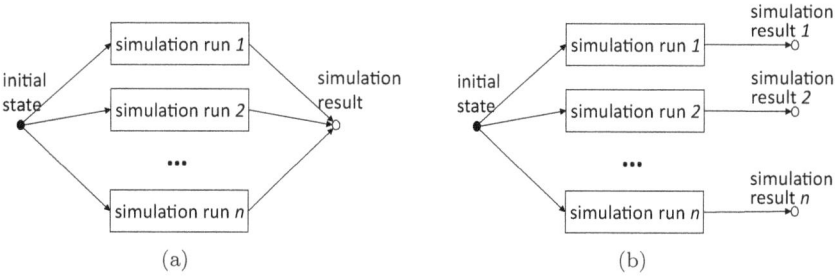

Figure 3.3. Simulation results of deterministic model and stochastic model: (a) deterministic model — different simulation runs with the same initial state and input trajectory would generate exactly the same result; (b) stochastic model — different simulation runs with the same initial state and input trajectory would generate different simulation results. Note that for simplicity the input trajectory is not shown in the figure.

numbers is repeatable if using the same starting value (also called the *seed*). This feature is useful when there is a need to repeat simulation runs for stochastic models, e.g., to compare results for different experiment scenarios or to repeat a problem when debugging a simulation program.

3.2 A Structural View of Simulation Model and Simulator

To help describe the various simulation models and their simulation algorithms, this section presents, in an informal manner, a structural view of simulation model and simulator. The structural view will then be used to guide the description of the various simulation models and their simulation algorithms.

For the purpose of this discussion, we view a simulation model as a component:

$$\text{Simulation Model} = < u, y, x, \delta_t, \lambda_t >$$

that has internal state x, responds to external input u, and generates output y. The dynamic behavior of the model is governed by the state transition function δ_t and the output function λ_t. δ_t defines how the model updates its state x over time t due to its internal dynamics and the external influence from input u. λ_t defines how the model

Figure 3.4. Simulation model and simulator — a structural view. (a) A general structure for simulation model and (b) separation of model and simulator.

generates output y based on its state x (here, we consider the Moore-type system, where output is determined by the state). The subscript t in δ_t and λ_t indicates the explicit role that time plays in the model behavior. Figure 3.4(a) shows the general structure of a simulation model.

The above structure does not pose any restrictions on how the state transition function and output function are defined. This is intentional because different models define them differently. The structure does not aim to be a complete specification for simulation models either (for example, how time advances is not defined). Its goal is to help describe the various simulation models in a structured way.

It is desirable to separate the simulation model from the simulation algorithm (i.e., the *simulator*) that executes the model. The clear separation between a model and its simulator, as shown in Figure 3.4(b), conforms to the modeling and simulation framework described in Section 2.2. The separation brings about several advantages from a practical point of view. First, a model and its simulator can be verified separately. This reduces the complexity of verification that is crucial for ensuring simulation correctness. Second, the modular design makes it easier to support reuse: The same simulator can simulate different models; and the same model may be simulated by different simulators when there is a need to execute the model in different ways. For example, a model may be simulated on a single computer (referred to as *centralized simulation*) or in a distributed environment using multiple computers (referred to as *distributed simulation*). With the separation of model and simulator, the switch from a centralized simulation to a distributed simulation does not need to modify the simulation model — it can be achieved by replacing the centralized simulator with a distributed simulator that implements a parallel and distributed simulation algorithm.

A simulation model may be simulated in multiple modes. To describe these modes, we first differentiate several notions of *time* that are useful for discussing simulation execution (Fujimoto, 2000):

- *Physical time* refers to time in the physical system (i.e., the real system to be simulated). Physical time is measured by the ticks of physical clocks. For example, a wildfire that happened from noon, November 8, 2018 to noon, November 25, 2018
- *Simulation time* is also called *logic time*. Simulation time is representation of physical time within a simulation model. It is implemented as a time variable with floating point values.
- *Wall-clock time* refers to time during the execution of a simulation model, measured by the ticks of physical clocks. For example, a simulation run that started at 10:00 a.m. on May 23, 2022 and ended at 10:10 a.m. on May 23, 2022.

Depending on whether or not a simulation run is synchronized with the wall-clock time, the following two modes of simulation execution can be defined:

- *As-fast-as-possible simulation*: The simulation runs as fast as it can. There is no fixed relationship between advances in simulation time and advances in wall-clock time.
- *Real-time simulation*: The execution of the simulation is synchronized with the wall-clock time. Each advance in simulation time is paced to occur in synchrony with an equivalent advance in the wall-clock time.

The two modes of simulation are employed in different use cases. Typically, when simulations are used for system analysis and prediction, they need to run in the as-fast-as-possible mode so that the analysis/prediction results can be obtained in a timely manner. The real-time simulation mode is often used in simulation-based virtual environments, e.g., to support human-in-the-loop training and gaming. In these applications, synchronizing the simulation time with the wall-clock time is important so that the human players perceive the generated scenarios to be "realistic."

3.3   Continuous Model and Simulation Algorithm

Continuous simulation models use continuous state variables to model systems. These models are generally specified by a set of *differential equations*, each of which defines the rate of change (i.e., the derivative) for one of the state variables. The rate of change at a time is a function of all the state variables and input variables at that time. The full set of differential equations form the state transition function δ_t of a continuous model.

Let x_1, x_2, ..., x_n be the state variables and u_1, u_2, ..., u_l be the input variables, where n and l are integers describing the dimensions of the state vector and input vector, respectively. Then, the state transition function δ_t of a continuous model is formed by a set of first-order differential equations:

$$dx_1(t)/dt = f_1(x_1(t), x_2(t), \ldots, x_n(t), u_1(t), u_2(t), \ldots, u_l(t)),$$
$$dx_2(t)/dt = f_2(x_1(t), x_2(t), \ldots, x_n(t), u_1(t), u_2(t), \ldots, u_l(t)), \quad (3.1)$$
$$\vdots$$
$$dx_n(t)/dt = f_n(x_1(t), x_2(t), \ldots, x_n(t), u_1(t), u_2(t), \ldots, u_l(t)),$$

where t is time and $f_i(\ ), i = 1, \ldots, n$ are the functions defining the derivatives for state variables $x_i, i = 1, \ldots, n$. These functions have the state vector and input vector as arguments.

The outputs at a time depend on the state value at that time. Let y_1, y_2, ..., y_m be the output variables of the continuous model, where m is the dimension of the output vector. The output function λ_t is formed by the set of equations

$$y_1(t) = g_1(x_1(t), x_2(t), \ldots, x_n(t)),$$
$$y_2(t) = g_2(x_1(t), x_2(t), \ldots, x_n(t)), \quad (3.2)$$
$$\vdots$$
$$y_m(t) = g_k(x_1(t), x_2(t), \ldots, x_n(t)),$$

where $g_i(\ ), i = 1, \ldots, m$ are the output functions for the output variables $y_i, i = 1, \ldots, m$, which have the state variables as arguments.

The set of differential equations in Equation (3.1) provides the functions for computing the derivatives for all the state variables. Starting from the initial values of the state variables, one can use these functions to compute the derivatives and then apply integration to obtain new values for the state variables. Doing this in an iterative way, in theory, one can precisely calculate the state values at any time in the future. Nevertheless, such a calculation requires integration over a continuous time base. To simulate a continuous model on a digital computer, it is necessary to define a next computation instant and a nonzero time interval (called a *time step*). In this way, the continuous functions are divided into small time intervals, and then, a *numerical integration method* can be employed to approximate the integral.

We consider the simplest integration method, generally known as the *Euler method*. The idea underlying the Euler method is that a derivative of a variable $x(t)$ at time t is the limit of the difference quotient at that time as the increment Δt approaches 0. In mathematical terms,

$$\frac{dx(t)}{dt} = \lim_{\Delta t \to 0} \frac{x(t + \Delta t) - x(t)}{\Delta t}. \tag{3.3}$$

Thus, for a small enough Δt, one can use the approximation

$$x(t + \Delta t) \approx x(t) + \Delta t \cdot \frac{dx(t)}{dt}. \tag{3.4}$$

The Euler method is straightforward to apply. However, it requires the time interval Δt to be sufficiently small in order to obtain accurate results. A smaller time interval means more iterations of calculation and hence a higher computation cost for a given length of run. Many other numerical integration methods, such as the Runge–Kutta methods, exist that often show better speed/accuracy trade-offs than the simple Euler method.

Algorithm 3.1 shows a general simulation algorithm for continuous simulation, which uses the Euler method as the integration method. To run the simulation, one needs to provide a time step Δt as well as the initial values for the state variables.

Algorithm 3.1. Continuous Simulation Algorithm.

t: simulation clock, starting from 0
Δt: time interval
T_f: simulation end time
$x_1(0), x_2(0), \ldots, x_n(0)$: initial state
$u_1(t), u_2(t), \ldots, u_l(t)$: external inputs at any time t (assumed to be known)

$t = 0$;
while $t \leq T_f$
    compute output $y_1(t), y_2(t), \ldots, y_m(t)$ using Equation (3.2);
    compute $\frac{dx_1(t)}{dt}, \frac{dx_2(t)}{dt}, \ldots, \frac{dx_n(t)}{dt}$ using Equation (3.1);
    compute $x_1(t + \Delta t), x_2(t + \Delta t), \ldots, x_n(t + \Delta t)$ using
        Equation (3.4);
    $t = t + \Delta t$; //update the simulation clock
endwhile

Simulation algorithms using other numerical integration methods would have a similar structure as the one shown above, with the difference that the step to compute $x_1(t + \Delta t), x_2(t + \Delta t), \ldots,$ $x_n(t + \Delta t)$ would use different integration methods.

3.3.1 *Continuous simulation example*

We consider the Lorenz system as an example of continuous simulation. The state transition function of the Lorenz system is defined by three ordinary differential equations known as the Lorenz equations:

$$\frac{dx}{dt} = \sigma(y - x),$$

$$\frac{dy}{dt} = x(\rho - z) - y, \tag{3.5}$$

$$\frac{dz}{dt} = xy - \beta z,$$

where x, y, and z are the three state variables (corresponding to x_1, x_2, and x_3 in Equation (3.1)). The σ, ρ, and β are the parameters

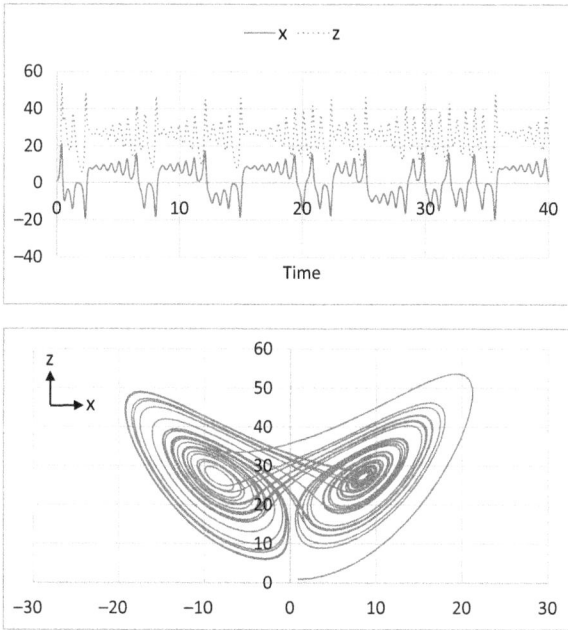

Figure 3.5. Continuous simulation example — chaotic behavior of the Lorenz system: The top figure shows how x and z change over time; the bottom figure shows their trajectory in the x–z plane.

of the model and have the following values: $\sigma = 10$, $\rho = 28$, and $\beta = 8/3$ (these parameter values make the Lorenz system have chaotic behavior). We are interested in seeing how the state variables change over time, thus we define the output variables to be the same as the state variables x, y, and z. The output function is omitted here due to its simplicity.

We use Algorithm 3.1 to simulate the Lorenz system. The simulation starts from initial values $x = 1.0$, $y = 1.0$, and $z = 1.0$ and uses a time step of $\Delta t = 0.01$. The simulation runs for 40.0 time units. Figure 3.5 shows the simulation results for the state variables x and z.

3.4 Discrete Time Model and Simulation Algorithm

A discrete time model updates its state at discrete time points called *time steps*. At each time step, the model is in a particular state, and it defines what the state at the next time step will be. The next state usually depends on the current state and current environmental influences (i.e., the external inputs). The notion of a time step is essential for discrete time models. For different applications, the unit of a time step may vary — it could be one second, one day, or one year. Sometimes, the time step is used to define an artificial clock to study how a system's state evolves from one "observation" instant to the next (as in the cellular automata example to be described in the following). In this case, the actual duration of the time unit is not important unless there is a need to align the model with scenarios in the real world.

The state transition function δ_t of a discrete time model defines how its state changes over time. It is generally specified in a difference equation form, as shown in the following:

$$x(t + 1) = \delta(x(t), u(t)), \tag{3.6}$$

where $x(t)$ and $x(t+1)$ are the state vectors at time steps t and $t+1$, respectively, and $u(t)$ is the input vector at step t. The time step t can only take integer values (starting from 0) because the time base of discrete time models needs to be isomorphic to the integers. The state vector x may include a single state variable or multiple state variables. The function $\delta()$ specifies how the next state is determined based on the current state and the current input. $\delta()$ may be defined in different ways. For example, it may be defined by a mapping table (as in the cellular automata example described in the following), by a set of rules (as in the Game of Life example described in Section 2.1), by a mathematical equation, or by a complex decision-making model (e.g., those used by autonomous robots).

The output of a discrete time model can be specified as

$$y(t) = \lambda(x(t)), \tag{3.7}$$

where $y(t)$ is the output vector, whose value depends on the state $x(t)$ at step t, and $\lambda()$ is the output function defining the mapping from $x(t)$ to $y(t)$.

A general discrete time simulation algorithm is provided in Algorithm 3.2. The algorithm proceeds in a stepwise fashion starting from step 0. In each step, the algorithm invokes the output function to compute the output of that step and then computes the next state using the state transition function. It then updates the simulation clock to the next time step and starts a new iteration.

Algorithm 3.2. Discrete Time Simulation Algorithm.

t: time step (i.e., simulation clock), starting from 0
T_f: simulation end time
$x(0)$: initial state
$u(0), u(1), \ldots, u(T_f)$: input trajectory (assumed to be known)

$t = 0$;
while $t \leq T_f$
    $y(t) = \lambda(x(t))$; // compute output
    $x(t+1) = \delta(x(t), u(t))$; // compute next state
    $t = t + 1$; //update the simulation clock
endwhile

3.4.1 *Discrete time simulation example*

We present the example of a one-dimensional cellular automaton, which is a classic example of discrete time simulation. In one-dimensional cellular automata, cells are all identical and are connected to form a one-dimensional grid, where each cell (except for the left and right end cells) has a left neighbor and a right neighbor. A cell and its two neighbors form a neighborhood of three cells. A cell can only be in two possible states: 0 or 1. Time is discrete, and the state of a cell at time $t+1$ is a function of the neighborhood cells' states at time t. This function is called the *transition rule* of the cellular automata. With each cell having two states and a neighborhood including three cells, there are $2^3 = 8$ possible patterns for the neighborhood state. For example, the pattern 000 means all three cells have a state value of 0, and the pattern 011 means the left, center, and right cells' states are 0, 1, and 1, respectively. For each pattern, a transition rule can be defined to make the center cell's next state be 0 or 1. This means, together, there exist $2^8 = 256$ possible transition rules for defining a center cell's next state.

Walfram used a standard naming convention for each of the 256 transition rules and systematically studied them (Walfram, 2002).

Table 3.1.   Rule 30 cellular automaton.

Current neighborhood state	111	110	101	100	011	010	001	000
Next state for the center cell	0	0	0	1	1	1	1	0

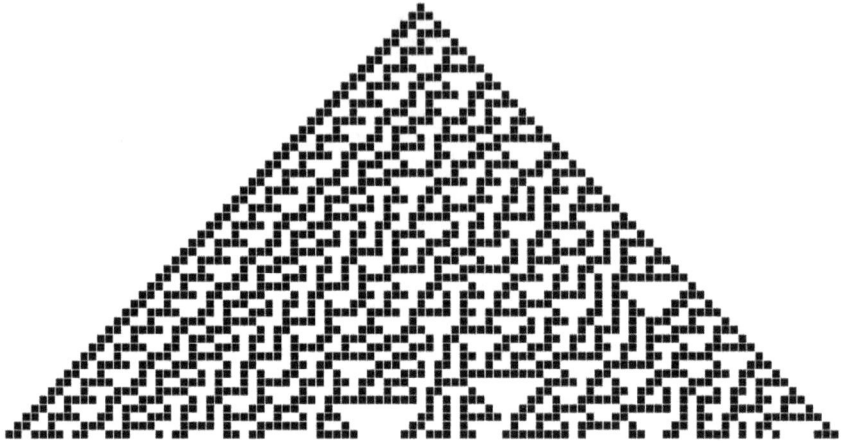

Figure 3.6.   Discrete time simulation example — the Rule 30 cellular automaton.

Table 3.1 shows the transition rule of the "rule 30 cellular automaton" (the number 30 is used because the binary code 00011110 represents the decimal value of 30). This table defines the state transition function δ_t for the rule 30 cellular automaton.

Given a specific rule and an initial state for a one-dimensional cellular automaton, one can use the discrete time simulation algorithm (Algorithm 3.2) to simulate how the cells change their states over time. In particular, when applied to an initial state of a single 1 surrounded by all 0s, the rule 30 cellular automaton generates the pattern as shown in Figure 3.6. The figure shows the results of 100 steps of the simulation. Each row in the figure represents a generation in the history of the automaton computed in one step of the discrete time simulation. The first generation is displayed on the top, and for each new generation, a new row is added below. Each cell is colored white for 0 and black for 1. Rule 30 exhibits what Wolfram calls *class 3 behavior*, meaning even simple input patterns lead to chaotic, seemingly random, histories.

3.5 Discrete Event Model and Simulation Algorithm

A discrete event simulation model models the operation of a system as a sequence of events in time. Events can be caused by the external environment. The occurrence of such external events is not under the control of the system itself. Alternatively, the system may schedule its own events to occur, which are called internal (time) events. When an event occurs, it can trigger a state change and result in the scheduling of new events and/or canceling old events. Between consecutive events, no change in the system is assumed to occur. Thus, a discrete event simulation can directly jump to the time of the next event.

Discrete event simulation models have been defined in a variety of ways. Formally, all discrete event simulation models can be specified using the DEVS formalism (Zeigler *et al.*, 2000). The material in this section focuses on discrete event simulation based on the DEVS formalism.

3.5.1 *Discrete Event System Specification (DEVS) model*

A DEVS model is a discrete event simulation model based on the DEVS formalism. DEVS differentiates a basic component from a composite component that is composed of other components. The basic component is called the *atomic model*, and the composite component is called the *coupled model*.

An atomic model has the structure

$$\text{DEVS}_{atomic} = < X, Y, S, \delta_{ext}, \delta_{int}, \delta_{con}, \lambda, ta >,$$

where
    X is the set of input values;
    Y is the set of output values;
    S is the set of states;
    $\delta_{int} : S \rightarrow S$ is the internal transition function, and
    $\delta_{ext} : Q \times X^b \rightarrow S$ is the external transition function, where
        $Q = \{(s, e) | s \in S, 0 \leq e \leq ta(s)\}$ is the *total state* set,
        e is the *time elapsed* since the last state transition,
        X^b denotes the collection of bags over X
        (sets in which some elements may occur more than once);

$\delta_{con} : S \times X^b \to S$ is the confluent transition function;
$\lambda : S \to Y^b$ is the output function;
$ta : S \to R^+_{0 \cup \infty}$ is the *time advance* function.

The input set X defines all possible inputs the atomic model may receive. The output set Y consists of all possible outputs the atomic model may send out. The set of states that the model can be in is denoted by S. Each state has an associated time duration that is defined by the time advance function ta. The state transition functions δ_{int}, δ_{ext}, and δ_{con} define how the model changes state under different conditions: δ_{int} specifies the state change in response to internal (time) events; δ_{ext} specifies the state change in response to external (input) events; and δ_{con} specifies the state change when both internal and external events occur at the same time. The output function λ defines how the model generates output. In reference to the structural view of simulation model (Figure 3.4(a)), one can see that δ_{int}, δ_{ext}, δ_{con}, and ta together define the state transition function δ_t of the model, and the output function λ_t is defined by λ as described above.

Figure 3.7 illustrates how the different elements of an atomic model work together to produce dynamic behaviors. At any time, the model is at some state s $(s \in S)$ and is scheduled to stay in that state for a period of time, the duration of which is defined by the

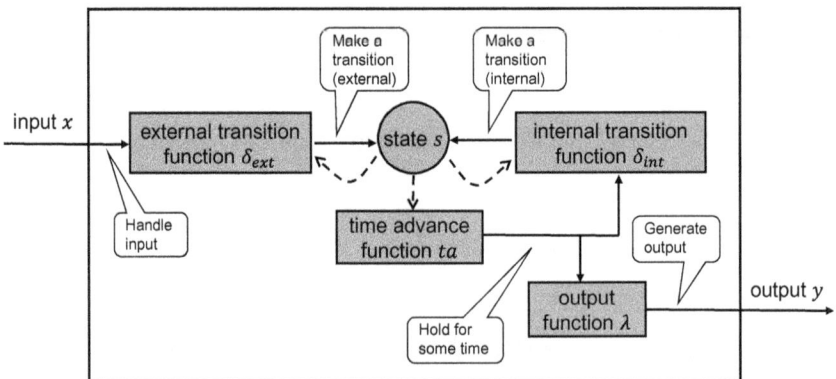

Figure 3.7.   How a DEVS atomic model works.

time advance function ta. This duration is called *sigma*, which represents the remaining time for the model to stay in a state. If there is no external input interrupting the model, its *sigma* will eventually expire. The expiration of *sigma* is called an *internal (time) event*. When that happens, the output function λ is first invoked to allow the model to generate outputs y ($y \in Y$). Then, the internal transition function δ_{int} is invoked to allow the model to transition to a new state. The new state would have its own *sigma*, indicating the time duration the model schedules to stay in the new state. A model may also receive an external input x ($x \in X$) before its *sigma* expires. This is considered an *external event*, which triggers the external transition function δ_{ext} to allow the model to respond to the external input immediately. δ_{ext} checks the model's current state s, the *elapsed time* e in the current state, and the input x to determine a new state (and a new *sigma* associated with the state). The confluent transition function δ_{con} deals with the special condition when external and internal events happen at the same time — it allows the model to decide how to handle collisions of internal and external events.

A coupled model is a composite model that is formed by coupling (i.e., connecting) several component models together. DEVS models are *closed under coupling*, meaning that a coupled model is equivalent to an atomic model and thus can be used as a component in a larger coupled model. This gives rise to hierarchical model construction, where large models are composed of component models in a hierarchical way. The modular modeling approach of DEVS has several advantages, including (1) reducing the modeling complexity because a large system can be broken down into manageable pieces to be modeled separately and then composed together; (2) supporting model reuse because component models can be tested and stored individually and then reused by coupling them with other components.

A coupled model contains the following information: the set of inputs, the set of outputs, the set of component models, and the coupling specification that defines how models influence each other. Typically, coupling is specified through models' input/output ports. Through a coupling from one model's output port to another model's input port, a model can send messages to another model. The arrivals of such messages represent external inputs for the receiving model.

A formal specification of the DEVS coupled model is omitted here. Readers can refer to the literature (e.g., Zeigler *et al.*, 2000) and use the example in Section 3.5.3 to understand how DEVS coupled models work.

3.5.2 *Discrete event simulation algorithm*

A discrete event simulation proceeds in an iterative way, driven by the sequential handling of events. A discrete event simulation algorithm maintains a list of events sorted by their scheduled event time, among which the event with the smallest time is called the *next event*. In each iteration, the simulation clock advances to the next event time, and the next event is processed. The processing of the next event can result in new events being scheduled or existing events being canceled. The event list is then updated accordingly, and a new simulation iteration starts. It is important to note that in discrete event simulation, the time intervals between events are irregular, and the simulation clock "jumps" to a new next event time in each iteration. This is different from discrete time simulation, where the simulation clock advances according to a fixed time step.

Based on the general idea of discrete event simulation described above, Algorithm 3.3 shows a simulation algorithm for simulating DEVS coupled models. A DEVS coupled model includes component models. In this algorithm, a *coordinator* is used to control the main simulation routine for the coupled model, while for each component model, there is a corresponding simulator that works with the coordinator and keeps track of the component model's local next event time. These simulators are called *component simulators* to differentiate them from the coordinator. Algorithm 3.3 (expressed in pseudocode) shows the simulation routine of the coordinator. For simplicity, we omit the pseudocode for the component simulators but provide a description of how they work with the coordinator.

In the algorithm, the coordinator uses two variables tL and tN to keep track of the global *last event time* and *next event time*, respectively. It maintains an event list (denoted by *eventList*) that keeps track of each component simulator's local next event time, denoted by tN_{local}. The *eventList* is sorted based on component simulators' tN_{local}. The smallest tN_{local} becomes the global tN in each iteration.

Algorithm 3.3. DEVS Simulation Algorithm (the coordinator).

tL: time of last event
tN: time of next event
T_f: simulation end time

eventList: an event list storing the pairs: (*simulator*, tN_{local})
imminents: set of simulators having the smallest next event time
receivers: set of simulators receiving inputs

// initialize all component models and the *eventList*
simulators.tellAll("initialize");
$tL = 0$;
$tN = eventList.getMin()$; // get the smallest tN_{local}
while $tN \leq T_f$

  imminents=eventList.getImms(tN); //get the *imminents* set
  imminents.tellAll("computeOutput",tN); // compute outputs
  //send outputs based on the couplings
  imminents.tellAll("sendOutput");
  imminents.add(receivers); // add *receivers* to the *imminents*
  //apply the state transition functions
  imminents.tellAll("applyDeltFunction",tN);
  imminents.tellAll("updateEventList"); // update the *eventList*
  // prepare for the next iteration
  $tL = tN$;
  $tN = eventList.getMin()$; // update the simulation clock
endwhile

In the beginning of the simulation, the coordinator asks all simulators to initialize their models and to populate the *eventList* with their tN_{local} information. The coordinator then obtains the global tN by executing the *eventListgetMin()* method. With the global tN at hand, the coordinator and the component simulators work together based on the following *simulation protocol* in each simulation iteration:

1. The coordinator gets the *imminents* from the *eventList*, which are the simulators having the smallest tN_{local}. There may be multiple simulators having the smallest tN_{local}, and all of them are added to the *imminents* set.

2. The coordinator asks all the simulators in the *imminents* set to invoke the output function of its model to compute the output messages.
3. The coordinator asks all the simulators in the *imminents* set to send the outputs. Each simulator checks its model's coupling information and puts its model's output message (if it is not empty) to the corresponding destination simulators. The output messages become the input messages of the destination simulators. The set of the destination simulators are called *receivers*. The *receivers*, like the *imminents*, need to invoke their models' state transition functions.
4. The coordinator adds the *receivers* to the *imminents* set by executing *imminentsadd(receivers)*.
5. The coordinator asks all the simulators in the *imminents* set to apply the state transition functions using the global tN as a parameter. Each simulator reacts to this request as follows:

   (a) If its tN_{local} equals to tN and its input message is empty, then it invokes its model's internal transition function δ_{int}.
   (b) If its tN_{local} equals to tN and its input message is not empty, then it invokes its model's confluence transition function δ_{con}.
   (c) If its tN_{local} does no equal to tN and its input message is not empty, then it invokes its model's external transition function δ_{ext}.

6. The coordinator asks all the simulators in the *imminents* set to update the *eventList* based on their new tN_{local} resulting from the execution of the state transition functions.
7. Then, the coordinator updates the global tL, obtains a new global tN from the *eventList*, and starts a new iteration.

The above description assumes that all component models are atomic models. When a component model is a coupled model, the simulator for that coupled model is set up in a similar way as a coordinator, which works with its own component simulators to get the next event time and to pass the corresponding requests from the parent coordinator to its component simulators. This can be set up in a hierarchical way to simulate coupled models with hierarchical structures.

3.5.3 Discrete event simulation example

We describe a car wash system as an example of discrete event simulation. The car wash system is a DEVS coupled model (named as *CarWashSys*) composed of three atomic models described as follows:

- *CarWashCenter*: This model models the behavior of a car wash center. The model starts in an *idle* state. When a car arrives, it becomes *busy* washing the car. It takes a fixed amount of time (8 time units in this example) to wash a car. Afterward, the finished car leaves, and the model goes back to being *idle*. The model rejects any incoming cars when it is *busy*.
- *CarGenerator*: This model generates car jobs for the *CarWashCenter* using a fixed inter-arriving time (6 time units in this example). The fixed inter-arriving time can be easily replaced by other distributions, such as exponential distribution.
- *Transducer*: This model monitors the performance of the *CarWashCenter* within a predefined time window (40 time units in this example). The model monitors the number of cars arrived at the *CarWashCenter* and the number of cars finished by the *CarWashCenter*. At the end of the time window, it sends a signal to stop the *CarGenerator* and the *CarWashCenter*.

Figure 3.8 shows the three models and their couplings. In the model, the car jobs generated from the *CarGenerator* are sent to the *CarWashCenter* to be processed. These jobs are also sent to the

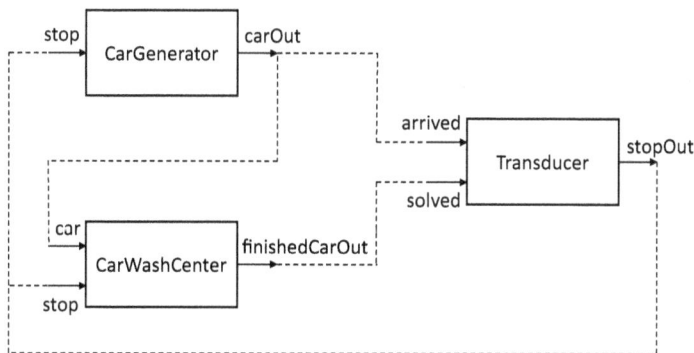

Figure 3.8. The *CarWashSys* coupled model: The model is composed of three atomic models; couplings among models' input and output ports are depicted as dashed lines.

arrived port of the *Transducer* to allow it to keep track of the number of cars arrived at the *CarWashCenter*. The finished cars from the *CarWashCenter* are sent to the *solved* port of the *Transducer* to allow it to keep track of the number of finished jobs. The *Transducer* is coupled to the *stop* input ports of both the *CarGenerator* and *CarWashCenter*. This allows the *Transducer* to stop the two at the end of the monitoring window.

The pseudocode of the three atomic models are given in Pseudocodes 3.1–3.3, where *initialize()* is the initialization function defining the initial state and *deltext()*, *deltint()*, and *out()* are the external transition function (δ_{ext}), internal transition function (δ_{int}), and output function (λ), respectively. In the code, the *holdIn()* method sets a model's state and associated time advance value (i.e., *sigma*). For example, *holdIn("active",6)* means setting a model's state to *active* and *sigma* to 6. The confluent transition function δ_{con} follows a default implementation that first invokes δ_{int} and then δ_{ext}. δ_{con} is omitted in the following pseudocodes. Also note that *deltext()* takes the elapsed time as one of the parameters. In this example, the three atomic models' state transitions do not use the elapsed time information. Thus, the elapsed time is not explicitly shown in the pseudocodes.

Pseudocode 3.1. The *CarGenerator* model.

```
inter_gen_time = 6; //inter generation time
initialize():  //initialization function
    holdIn("active", inter_gen_time);
deltext(): //external transition function δext
    if(receives message on "stop" port)
        holdIn("passive", infinity);
deltint(): //internal transition function δint
    if(phase is "active")
        holdIn("active", inter_gen_time);
out(): //output function λ
    if(phase is "active")
        sendOutputThroughPort("carOut");
```

Pseudocode 3.2. *The CarWashCenter* model.

```
car_wash_time = 8; //car wash time
initialize():  //initialization function
    holdIn("idle", infinity);
deltext(): //external transition function δext
    if(receives message on "car" port && phase is "idle")
        holdIn("busy", car_wash_time);
    if(receives message on "stop" port)
        holdIn("passive", infinity);
deltint(): //internal transition function δint
    if(phase is "busy")
        holdIn("idle", infinity);
out(): //output function λ
    if(phase is "busy")
        sendOutputThroughPort("finishedCarOut");
```

Pseudocode 3.3. *The Transducer* model.

```
monitor_time_window = 40; //the monitor time window
num_arrived_car = 0; //number of cars arrived at CarWashCenter
num_finished_car = 0; //number of cars finished by CarWashCenter
initialize():  //initialization function
    holdIn("active", monitor_time_window);
deltext(): //external transition function δext
    if(receives message on "arrived" port)
        num_arrived_car++; //increase num_arrived_car by 1
    if(receives message on "solved" port)
        num_finished_car++; //increase num_finished_car by 1
deltint(): //internal transition function δint
    if(phase is "active"){
        print num_arrived_car and num_finished_car result;
        holdIn("passive", infinity);
    }
out(): //output function λ
    if(phase is "active")
        sendOutputThroughPort("stopOut");
```

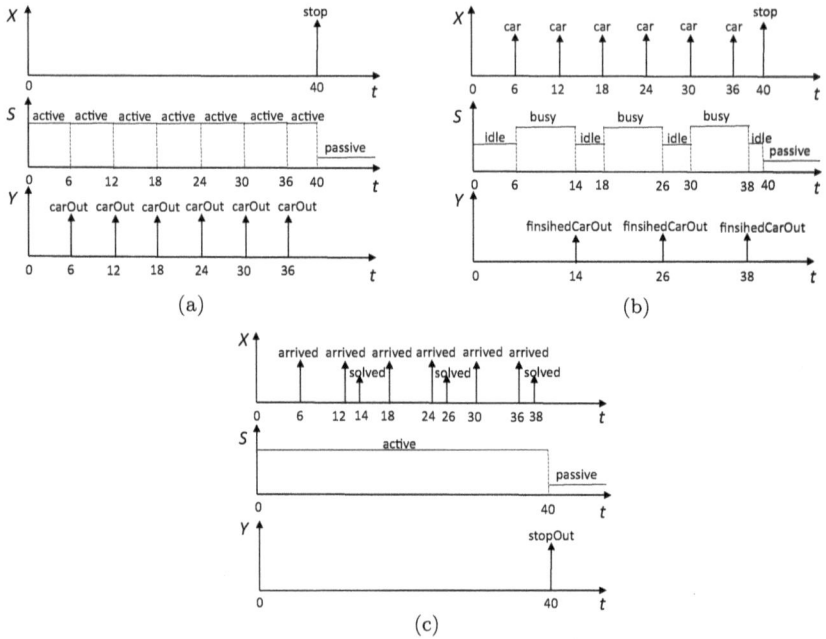

Figure 3.9.   Simulation results: input, state, and output trajectories of the three models. (a) *CarGenerator* model, (b) *CarWashCenter* model and (c) *Transducer* model. X — input trajectory, S — state trajectory and Y — output trajectory.

Figure 3.9 shows the input trajectory, state trajectory (only the discrete phases, such as *idle*, *busy*, and *passive*, are shown), and output trajectory of the three atomic models when simulating the car wash system. As can be seen, the *CarGenerator* is always *active* and keeps generating cars every 6 time units until it receives the *stop* message (from the *Transducer* model) at time 40. The *CarWashCenter* alternates between *idle* and *busy* until it receives the *stop* message at time 40 and becomes *passive*. The *Transducer* is *active* for 40 time units and then sends out the *stop* message to the *CarGenerator* and *CarWashCenter*.

The above car wash system model can be extended to account for more complex scenarios. For example, the *CarGenerator* can be extended to make the *inter-gen-time* follow a probability distribution (e.g., an exponential distribution), or to generate different types of cars. The *CarWashCenter* can be extended to have a queue to queue the jobs when it is busy, or to have multiple service lanes. The *Transducer* can be extended to calculate other results, such as average service time, or to stop monitoring if a certain condition (e.g., when the number of arrived cars reaches 100) is satisfied.

3.6 Agent-Based Model and Simulation Algorithm

3.6.1 *Agent-based model*

An agent-based model is a micro-level model that models a system as a collection of autonomous decision-making entities called *agents*. Each agent is an active computation component that pertains to a set of characteristics and has the capability of making its own decisions. Agent-based modeling has several advantages (Bonabeau, 2002), including: (1) it is able to capture emergent phenomena resulting from the interactions of individual entities; (2) it provides a natural way to model systems that are composed of "behavioral" entities; and (3) it is flexible to support agents with different variations, such as having different property values, behavior rules, and complexities.

Agent-based models have been defined in a variety of ways. Agents can be defined as system components with inputs, outputs, and state transition functions in a way similar as DEVS models described in the previous section. More often, agent-based models are defined based on the point of view of an "autonomous agent," where each agent is thought to follow a *sense–think–act* procedure to define its behavior. The *sense–think–act* procedure is invoked iteratively as time advances.

- **Sense:** The sense step allows an agent to obtain the necessary information pertaining to itself and its environment. The environment includes not only the physical or social space where the agent is situated but also other agents that influence this agent's behavior. Typically, an agent senses the environment within its local neighborhood. For different applications, the neighborhood may be defined differently. For example, a neighborhood may be defined based on the Euclidean distance centered around each agent or based on the social network connections of each agent.
- **Think:** The think step allows an agent to make decisions based on the sensed information. Depending on the complexity of the decision-making model, agents may be categorized as *simple agents*

or *complex agents*. Simple agents use relatively simple behavior rules to decide how they evolve and respond to environmental inputs. Complex agents use complex decision-making models, e.g., models involving learning, cognition, and/or optimization. Complex agents are often used to simulate systems in social and cognitive sciences, economics, and artificial societies.

- **Act:** The act step allows an agent to perform the actions associated with the decisions made in the think step. This is the last step of the sense–think–act cycle. The act step changes the agent's properties and/or the properties of the environment. The changed property values result in new situations for agents that will be sensed in the next sense–think–act cycle.

The *sense–think–act* procedure may not be strictly followed by all agent-based models. Sometimes, one or all of the sense, think, and act steps are so simple that there is no need to explicitly separate them when defining an agent. Consider the one-dimensional cellular automata described in Section 3.4, which can be considered an agent-based model where individual cells are agents. In this example, the sense step of a cell is to obtain its own state as well as its left and right cells' states; the think step is to apply the transition rule to compute the new state; and the act step is to update the new state. When cells' states are stored in an array that can be directly accessed and modified, all three steps can be realized by a single function that updates a cell's state based on the state transition table (Table 3.1).

From the *timing of change* point of view, an agent-based model can be implemented as a discrete time model or a discrete event model. In a discrete time implementation, agents update their states in a stepwise fashion based on a fixed time step. In each step, each agent goes through its own sense–think–act procedure to sense the environment, make a decision, and carry out the action. The state transition function of the overall model includes the state transitions of all the agents. In a discrete event implementation, agents' state updates are not synchronized by a fixed time step. Instead, each agent updates its state according to its own schedule (i.e., internal events) or in response to changes from the environment (i.e., external events). When an agent updates its state, it notifies its neighboring agents. Such notifications are external events of the neighboring agents. For agents that have continuous state variables, such as positions in a continuous Euclidean space, a *quantization* mechanism may be used to define a threshold of change, based on which events can be defined. An example of a discrete event implementation of an agent-based pedestrian crowd simulation can be found in the work of Qiu and Hu (2013).

3.6.2 *Agent-based simulation algorithm*

Simulations of agent-based models depend on the *timing of change* mechanisms of the models. For agent-based models that are implemented in a discrete time fashion, simulation would be carried out using a discrete time simulation algorithm. For agent-based models that are implemented in a discrete event fashion, simulation would be carried out using a discrete event simulation algorithm.

In the following, we assume a discrete time implementation and present a discrete time simulation algorithm for agent-based simulation. This algorithm follows the basic structure of the discrete time simulation algorithm described in Algorithm 3.2 but incorporates the *sense–think–act* procedure described above. A time step is used to advance the simulation clock in a stepwise fashion (the time step is assumed to be 1 in this algorithm). In each step, each agent goes through its sense–think–act procedure to compute and update its state. Note that a discrete time agent-based simulation needs to ensure that all agents' state updates are synchronized by the time steps. At a particular step, agents should only use the information of that step, as opposed to the information that is newly updated during the step. This means an agent should not commit the state change to the next time step until all other agents have finished computation for the current time step.

Algorithm 3.4 shows a general structure for a discrete time agent-based simulation. In the algorithm, $sense_i()$, $think_i()$, and $act_i()$ are steps of the sense–think–act procedure for agent $i, i = 0, 1, \ldots, N-1$, where N is the total number of agents. The algorithm uses a two-stage implementation to ensure that the agents' state updates are synchronized. In the first stage, all agents compute their new states and store them in a temporary state array called q_next. After all agents finish computation for the step, the second stage commits the state changes by assigning the state values stored in q_next to the state array that will be used in the next cycle.

Algorithm 3.4. Discrete Time Agent-based Simulation Algorithm.

t: simulation clock, starting from 0
T_f: simulation end time
$q_i(0)$: initial state of agent $i, i = 0, 1, \ldots, N - 1$
$q_i(t)$: state at time t for agent $i, i = 0, 1, \ldots, N - 1$
q_next_i: temporarily storing agent i's next state, $i = 0, 1, \ldots, N-1$

$t = 0$;
while $t \leq T_f$
    for each agent $i(i = 0, 1, \ldots, N - 1)$ in the agent list
      $sense_i(\ )$; //the sense step
      $think_i(\ )$; //the think step
      $q\_next_i = act_i(\ )$; //the act step
    endfor
    for each agent $i\ (i = 0, 1, \ldots, N - 1)$ in the agent list
      $q_i(t + 1) = q\_next_i$; //commit the state change
    endfor
    $t = t + 1$; //update the simulation clock
endwhile

3.6.3 *Agent-based simulation example*

We describe the Boids model (Reynolds, 1987) as an example of agent-based simulation. The Boids model was developed by Craig Reynolds in 1987 to simulate the flocking behavior of birds. The model has often been used to demonstrate emergent behavior that can arise from the interactions of individual agents that follow relatively simple rules.

The Boids model is an agent-based model that consists of a group of agents named *boids* (bird-like objects). Each boid has its own position, velocity, and orientation. It can sense other boids (also referred to as *flockmates*) within a relatively small neighborhood around itself. The neighborhood is characterized by a distance and an angle, as illustrated in Figure 3.10(a). This neighborhood defines the region in which flockmates influence a boid's steering. A boid's steering in each step is influenced by three rules (or behaviors), which

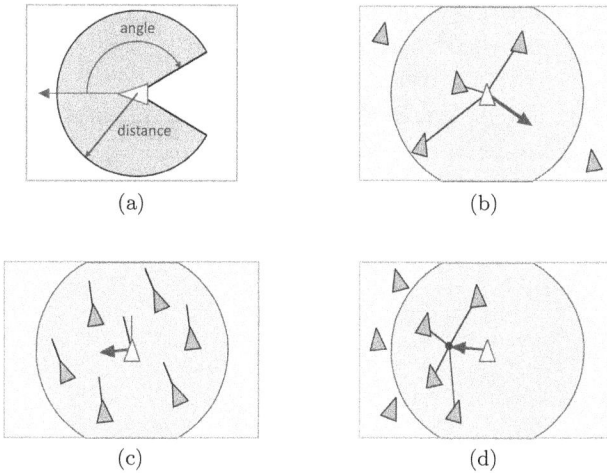

Figure 3.10. The Boids model. Reproduced with permission from Reynolds (1999). (a) a boid's neighborhood; (b) Separation: steer to avoid crowding local flockmates; (c) Alignment: steer toward the average heading of local flockmates (d) Cohesion: steer to move toward the average position of local flockmates.

are described as follows and illustrated in Figure 3.10(b)–(d). These behaviors are all based on information about local flockmates defined by the neighborhood:

- **Separation:** Steer to avoid crowding local flockmates. For this behavior, a boid gets the closest flockmate and steers to the opposite direction from the closest flockmate.
- **Alignment:** Steer toward the average heading of local flockmates. For this behavior, a boid gets the average heading direction of all flockmates within its neighborhood and steers to match its own heading with the average heading direction.
- **Cohesion:** Steer to move toward the average position (center of mass) of local flockmates. For this behavior, a boid computes the average position of all the flockmates within its neighborhood and steers toward it.

The three behaviors (separation, alignment, and cohesion) each produce a speed vector. These speed vectors are all added to the boid's current speed vector to generate a new speed vector. The boid then updates its position based on the new speed vector. The simulation runs in a stepwise fashion (a discrete time simulation). In each step, each boid computes its speed vector and updates its position as described above.

The boids model simulation is able to generate realistic-looking behaviors similar as those observed in a flock of birds in the real world. Unexpected behaviors, such as flocks splitting and reuniting after avoiding obstacles, are considered examples of emergent behavior. An animation of a boids model simulation can be found in Wikipedia (n.d.).

3.7   Cellular Space Model

Cellular space models originate from cellular automata. A cellular automaton is a discrete model of computation in which space and time are discretized and the state sets are discrete and finite. Cellular automata were originally introduced by von Neumann and Ulam as an idealization of biological self-production (Burks, 1970).

In cellular space models, space is discretized into individual cells, which are first-class components that have their own behavior. Cells are geometrically located on a one-, two-, or multidimensional grid and connected in a uniform way. The cells that influence a particular cell are called the *neighborhood* of the cell, which are often chosen to be the cells located nearest in the geometrical sense. Each cell models a local portion of the overall space, and a cell's behavior is influenced by its neighborhood cells as well as its own state. Cellular space models have the advantage that they discretize a space into space entities (i.e., cells) that have their own explicit behaviors. They are often used to model applications where space plays an essential role in the dynamics of the system, such as wildfire spread in forests, crowd movement in public spaces, and urban development in metropolitan areas.

Cellular space models are micro-level models because they model systems using individual cells. They are considered a special kind of agent-based model where agents are cells and agents' connections are defined by the neighborhood structure of cells. A cellular space model may also work with other agents, which are separate components from the cells and interact with the cells. For example, a wildfire suppression simulation may model the dynamic spread of a wildfire using a forest cellular space and model the fire suppression using firefighter agents that act on the forest cells.

Simulation of a cellular space model follows the same procedure as simulating an agent-based model. Specifically, for a cellular space model that is implemented in a discrete time fashion, a discrete time simulation algorithm (e.g., Algorithm 3.2) would be used. For a cellular space model that is implemented in a discrete event fashion, a discrete event simulation algorithm (e.g., Algorithm 3.3) would be used.

3.7.1 *Cellular space simulation example*

An example of the cellular space model for wildfire spread simulation will be presented in Chapter 8. This section describes another example that simulates an ecosystem made up of sheep and grass. A high-level description of this system is given below.

The *sheep–grass ecosystem* is modeled as a two-dimensional cellular space. Each cell has eight neighboring cells (the Moore neighborhood). The cellular space is wrapped so that the left and right boundaries and the top and bottom boundaries are connected. A cell can be an *empty* cell, a *grass* cell, or a *sheep* cell. Their behaviors are described as follows:

- An empty cell stays empty until it is occupied by grass (due to grass reproduction) or a sheep (due to sheep movement or reproduction).
- A grass cell lives forever until it is eaten by a sheep. A grass cell reproduces itself every $T_{\text{grassReproduce}}$ time units. When reproducing, a grass cell reproduces grass in a randomly selected adjacent cell that is empty. If there is no empty adjacent cell (i.e., all adjacent cells are either grass or sheep), the reproduction has no effect.
- A sheep cell has a sheep on it. A sheep has the following behavior:

  o A sheep constantly moves to eat grass. It takes $T_{\text{sheepMove}}$ time units for a sheep to move to a neighboring cell. When moving, a sheep moves to a randomly selected neighboring cell that has grass. If there is no grass cell in its neighborhood, a sheep moves to a randomly selected empty cell. If all adjacent cells are sheep, the movement has no effect (i.e., the sheep ends up at its original cell). After a sheep moves to a neighboring cell, the original sheep cell becomes an empty cell.

- ○ If a sheep fails to eat grass within $T_{\text{sheepLife}}$ time units, the sheep dies.
- ○ A sheep reproduces every $T_{\text{sheepReproduce}}$ time units. When reproducing, a sheep reproduces a sheep in a randomly chosen adjacent cell at the end of the $T_{\text{sheepReproduce}}$ time (as a result, the adjacent cell has a sheep on it). If all eight adjacent cells already have sheep, the reproduction has no effect.

The sheep–grass ecosystem described above is an example of the so-called *predator and prey* system. The dynamics of this system depends on the parameter values of $T_{\text{grassReproduce}}$, $T_{\text{sheepMove}}$, $T_{\text{sheepLife}}$, and $T_{\text{sheepReproduce}}$ as well as the initial condition of grass and sheep cells. With the right parameter values and initial condition, the populations of grass and sheep can oscillate in a cyclic fashion. A cycle can be explained in the following way: When there is lots of grass, sheep can easily survive and thus reproduce more and more sheep over time. The greater number of sheep would consume more grass and eventually make the grass population to decrease. When grass becomes rarer, more sheep begin to starve and die. This reduces the sheep population and grass consumption, and as a result, the grass can go back to rapidly reproducing. Then, the cycle repeats itself.

Figure 3.11 (left) shows a snapshot of a sheep–grass ecosystem simulation in a 40×40 cellular space. The cellular space model is implemented as a DEVS coupled model. It has the following parameter values: $T_{\text{grassReproduce}} = 1.0$, $T_{\text{sheepMove}} = 1.0$, $T_{\text{sheepLife}} = 3.0$, and $T_{\text{sheepReproduce}} = 5.0$. Figure 3.11 (right) shows the grass and sheep populations (i.e., number of cells) over 200 time units from a simulation run. One can see the cyclic oscillation of the grass and sheep populations. The cycles do not repeat themselves exactly due to the randomness and spatial heterogeneity of the grass and sheep cells in each cycle.

3.8 System Dynamics Model

System dynamics is a modeling paradigm where a complex system is modeled by the internal feedback loops and time delays that exist in the system to influence its behavior. System dynamics was originally developed by Forrester (1972) to help corporate managers to

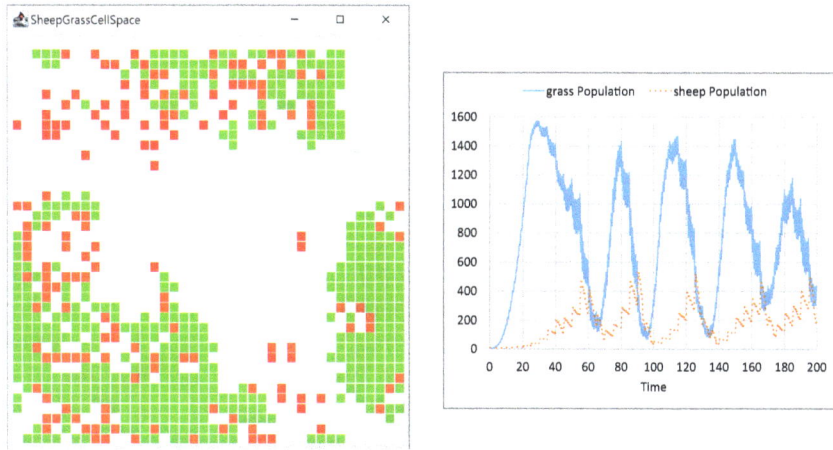

Figure 3.11. Cellular space modeling and simulation for the sheep–grass ecosystem: (left) a simulation snapshot showing grass cells (green), sheep cells (red), and empty cells (white); (right) the cyclic oscillation of sheep and grass populations over time.

better understand industrial processes. Since then, its application has expanded to many other fields, such as natural systems, business, economics, social systems, and public health systems. System dynamics models are macro-level models that adopt a top-down modeling approach, where a system is modeled as a whole with its internal structures.

System dynamics models use *stocks* and *flows* to model the constituent components and their interactions in a system. A *stock* represents a quantity that can accumulate over time; a *flow* represents the rate of change of a stock. Stocks can only be changed via flows — they are accumulated over time by inflows and/or depleted by outflows. Stocks have a certain value at each moment of time, while flows are measured over an interval of time. Figure 3.12 shows a stock and flow diagram for a savings account, where the savings account balance is the stock variable and the interest is the flow (inflow). In this example, we assume there is no withdrawal activity for the savings account and thus there is no outflow for the stock.

System dynamics models emphasize the feedback loops that may exist in a system. A feedback loop exists when some outputs of a system are routed back as inputs as part of a chain of cause and effect. If the tendency of the loop is to reinforce the inputs, the

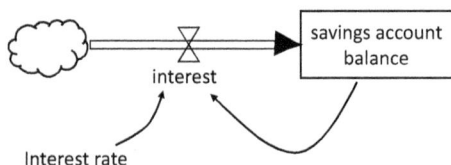

Figure 3.12.   Stock and flow diagram for a savings account.

loop is called a *positive* or *reinforcing feedback loop*; if the tendency
is to oppose the inputs, the loop is called a *negative* or *balancing
feedback loop*. The stock and flow diagram in Figure 3.12 includes
a positive feedback loop because the interest flow depends on the
account balance — more account balance generates more interest
flow. Due to this positive feedback loop, the account balance would
increase exponentially over time.

The stock and flow diagram provides a graphical representation of
a system dynamics model. Mathematically, a system dynamics model
is generally specified using a set of first-order differential equations.
Stocks are state variables. The derivative of a stock equals the total
effect of all its inflows and outflows. Following this principle, one can
convert a stock and flow diagram to the corresponding differential
equations. Consider the example shown in Figure 3.12. The following
differential equation can be derived from the stock and flow diagram:

$$\frac{dx}{dt} = rx, \tag{3.8}$$

where x is the savings account balance (i.e., the stock) and r is the
fixed interest rate.

It is important to note that while system dynamics models are
specified by differential equations that work with continuous state
variables, many system dynamics models deal with stock values that
are discrete in nature. For example, the balance in a savings account
would have a basic unit of 1 cent. Other examples include population dynamics, where the state variables are numbers of people,
which are integers by definition. For these applications, the state
variables are treated as continuous variables during a simulation,
but the final results would be rounded to the closest discrete values
that are meaningful for the applications.

The underlying procedure for simulating system dynamics models
is the same as that for simulating continuous simulation models.

Once the differential equations of a system dynamics model are defined, they can be simulated using the same simulation algorithm as Algorithm 3.1.

3.8.1 *System dynamics model example*

System dynamics can be used to model the SIR model for infectious disease spread simulation. The SIR model is an epidemiological model that computes the number of people infected with an infectious disease, such as COVID-19, in a closed population over time. The name of the model derives from the fact that it divides the population into three subgroups (or compartments): *Susceptible*, *Infected*, and *Recovered*, denoted by S, I, and R, respectively.

- S: The number of susceptible people. These are the individuals who have not caught the infectious disease yet. A susceptible individual may become infected when having contact with infected individuals.
- I: The number of people who have been infected (and not recovered yet). Infected individuals are capable of infecting susceptible individuals.
- R: The number of people who have been infected and recovered from the disease. Recovered individuals will not be infected by the disease again. Nor will they infect susceptible individuals.

Figure 3.13 shows the stock and flow diagram of the SIR model. There are three stocks in this model: S, I, and R. The stock S has no inflow; it is depleted as people become infected due to contact with infected individuals. The rate at which susceptible individuals become infected is called the *infection flow*. This flow depends on the number of susceptible individuals (S), the proportion of infected individuals (I/N, where N is the total population), and an effective transmission rate (β), which accounts for the transmissibility of the disease as well as the average number of contacts per individual. The stock I has one inflow which is the infection flow. It also has an outflow called the *recovery flow*, which specifies the rate of people recovering from the disease. The recovery flow is assumed to be proportional to the number of infectious individuals, i.e., γI, where γ is the recovery rate, defined as the inverse of the recovery duration of the disease.

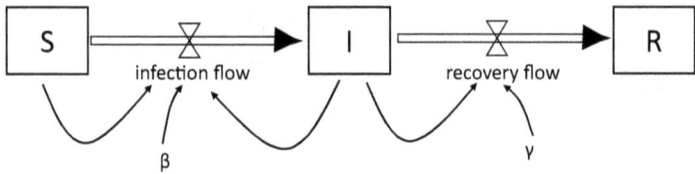

Figure 3.13.   Stock and flow diagram of the SIR model.

Based on the stock and flow diagram shown in Figure 3.13, the equations of the SIR model are

$$\frac{dS}{dt} = -\beta \left(\frac{I}{N} \right) S,$$

$$\frac{dI}{dt} = \beta \left(\frac{I}{N} \right) S - \gamma I, \tag{3.9}$$

$$\frac{dR}{dt} = \gamma I,$$

where $N = S + I + R$ is the total population. This is a constant because $\frac{dS}{dt} + \frac{dI}{dt} + \frac{dR}{dt} = 0$.

Figure 3.14 shows the simulation results of an SIR model. The model has the following initial states and model parameters: $S(0) = 10^6$, $I(0) = 10$, $R(0) = 0$, $\beta = 0.5$/day, and $\gamma = 0.1$/day. The simulation is carried out using Algorithm 3.1. As can be seen, at the beginning of the simulation, the numbers of infected people and recovered people arc small because there are few infected individuals and the disease spreads slowly. As more individuals become infected, they contribute to the spread and increase the infection flow. This results in an exponential increase in the number of infected people. As the number of people in the infected compartment increases, that in the susceptible compartment decreases and that in the recovered compartment increases. The number of infected people peaks at day 33 and then starts to decrease. By day 100, the number of infected people decrease to 1140; there are still 6,745 people in the susceptible state, and the majority of people are in the recovered compartment.

The basic SIR model has been extended in various ways to capture different aspects of disease spread and intervention. For example, a common extension is the so-called *susceptible–exposed–infectious–removed (SEIR)* model, which adds a new compartment of exposed

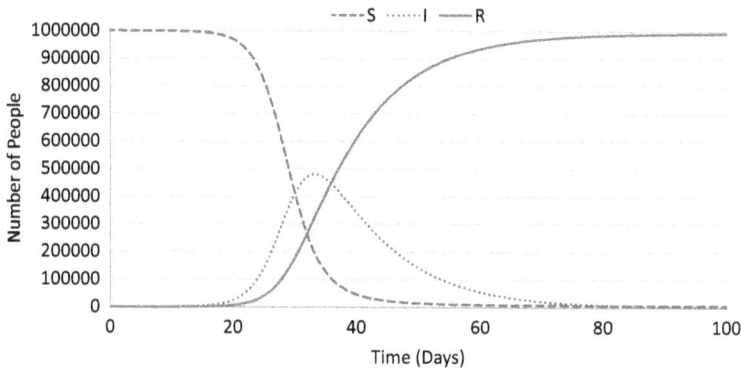

Figure 3.14. Simulation results of the SIR model.

people to the SIR model. The exposed compartment exists between the susceptible and the infectious compartments; it includes individuals who have been exposed to the infection but are not themselves infectious yet.

3.9 Sources

The descriptions of the three categories of models and simulation algorithms under the "timing of change" criterion: continuous model and simulation, discrete time model and simulation, and discrete event model and simulation, are influenced by the work of Zeigler *et al.* (2000). References used for describing other types of models are provided in the corresponding sections.

Chapter 4

Basic Probability Concepts

This chapter familiarizes readers with the basic probability concepts and notations used in this book. The content of this chapter is meant to serve as a quick reference for the concepts and notations that are useful for understanding the data assimilation topic to be described in Chapter 6. A more comprehensive description of the subject can be found in probability and statistics-related textbooks. Readers who are familiar with the related concepts may skip this chapter.

4.1   Random Variable and Probability Distribution

A *random variable* is a variable whose value cannot be defined with certainty. The value of a random variable is determined by the outcome of an experiment. Since different experiments can lead to different outcomes, we may assign probabilities to the possible values of a random variable. Let X denote a random variable and x denote a specific value that X might take on. If the space of all the values that X can take on is discrete, we denote the probability that the random variable X has a value x by

$$p(X = x). \tag{4.1}$$

Consider a coin flip as an example. The outcome of the coin flip is a random variable that can take the values of *head* or *tail*. For a fair coin, we have the probabilities $p(X = head) = 1/2$ and $p(X = tail) = 1/2$.

A note on notation: We use the capital letter X to represent a random variable and the small-case letter x to represent a specific value of X. For simplicity, in later descriptions, we use x to denote both the random variable and its realization unless there is a need to explicitly differentiate them.

A random variable can be discrete or continuous. A discrete random variable takes on only discrete values; a continuous random variable takes on values from a continuous range. The list of all possible values of a random variable and their associated probabilities are generally referred to as the *probability distribution* of the random variable.

A discrete random variable's probability distribution can be specified by its *probability mass function* (PMF). Let X be the discrete random variable and $R_X = \{x_1, x_2, x_3, \ldots\}$ be the range of all possible values that X can take on, then the PMF, denoted by $p(x)$, of X is defined by

$$p(x) = p(X = x) \quad \text{for } x \in R_X. \tag{4.2}$$

The PMF $p(x)$ is non-negative. Since X must take on one of the values from all the possible values, we have

$$\sum_{x_k \in R_X} p(x_k) = \sum_{x_k \in R_X} p(X = x_k) = 1. \tag{4.3}$$

A continuous random variable's probability distribution is often specified by its *probability density function* (PDF). Let X be a continuous random variable and $F(x) = p(X \leq x)$ be the *cumulative distribution function* that defines the probability that X takes on a value less than or equal to x. Then, the PDF (denoted by $f(x)$ for now) of X, defined for all real $x \in (-\infty, \infty)$, is a non-negative function that has the property that for any real number x,

$$F(x) = p(X \leq x) = \int_{-\infty}^{x} f(a)da. \tag{4.4}$$

Differentiating both sides of Equation (4.4) yields

$$f(x) = \frac{d}{dx}F(x). \tag{4.5}$$

That is, the density is the derivative of the cumulative distribution function. If a probability distribution has density $f(x)$, then the infinitesimal interval $(x, x + dx)$ has the probability $f(x)dx$. In other words, $f(x)$ is a measure of how likely it is that the random value X will be near x.

Just as discrete probability distributions always sum up to one, a PDF always integrates to one:

$$\int_{-\infty}^{\infty} f(x)dx = 1. \tag{4.6}$$

For a discrete random variable X with range R_X and PMF $p(x)$, a generalized PDF may be defined:

$$f(x) = \sum_{x_k \in R_X} p(x_k)\delta(x - x_k), \tag{4.7}$$

where $\delta(x)$ is the Dirac delta function whose value is zero everywhere except at zero and whose integral over the entire real line is equal to one.

The generalized PDF makes it convenient to use the same formulas to calculate the moments of discrete and continuous random variables. Hereinafter, for simplicity we use the PDF as a general term to refer to the probability distribution for both continuous and discrete random variables and denote it by $p(x)$.

A commonly used probability distribution is the one-dimensional *normal distribution* (also called the *Gaussian distribution*). A normal distribution is defined by its *mean* (denoted by μ) and *variance* (denoted by σ^2). The PDF for the normal distribution is given by

$$p(x) = \frac{1}{\sigma\sqrt{2\pi}}e^{-\frac{1}{2}\left(\frac{x-\mu}{\sigma}\right)^2}. \tag{4.8}$$

A normal distribution is often denoted by $x \sim N(\mu, \sigma^2)$, where x is the random variable, "$\sim$" means "distributed according to," N stands for the normal distribution, and μ and σ^2 are the mean and variance of the distribution, respectively.

4.2 Joint Distribution, Marginal Distribution, and Conditional Probability

The previous section focused on a single random variable. Often, we are interested in probability distributions concerning two or more random variables. The *joint distribution* of two random variables X and Y is given by

$$p(x, y) = p(X = x \text{ and } Y = y), \qquad (4.9)$$

which describes the probability of the event that the random variable X takes on the value x and that Y takes on the value y. When X and Y are *independent*, we have

$$p(x, y) = p(x)p(y). \qquad (4.10)$$

The joint distribution $p(x, y)$ is for both random variables X and Y. Given a joint distribution, sometimes we want to know the probability concerning a single variable, irrespective of the value of the other variable. For example, we may want to know the probability that X takes on the value of 4, i.e., $p(X = 4)$, regardless of what values that Y takes on. This type of probability that defines the probability of one (or a subset) of the random variables irrespective of the other ones is called the *marginal probability*. The distribution of the marginal probabilities is called the *marginal distribution*. Note that a marginal probability or marginal distribution is meaningful only within the context of multiple random variables.

Based on a joint distribution $p(x, y)$, the marginal distribution can be calculated as follows:

$$p(x) = \int p(x, y)dy, \qquad (4.11)$$

$$p(y) = \int p(x, y)dx. \qquad (4.12)$$

Often, we are also interested in calculating the probabilities when some partial information is known. In other words, the desired probabilities are conditioned on the partial information that is known. In probability theory, this is defined by the *conditional probability*. Suppose we already know that Y's value is y, and we want to know the probability that X's value is x conditioned on that fact.

This conditional probability is denoted by

$$p(x|y) = p(X = x|Y = y). \tag{4.13}$$

When $p(y) > 0$, the conditional probability $p(x|y)$ is calculated as

$$p(x|y) = \frac{p(x, y)}{p(y)}. \tag{4.14}$$

If X and Y are independent, we have

$$p(x|y) = \frac{p(x, y)}{p(y)} = \frac{p(x)p(y)}{p(y)} = p(x). \tag{4.15}$$

This means that when X and Y are independent, the known information of Y tells nothing about the value of X. In other words, there is no advantage of knowing Y if our interest is only on X.

Equation (4.14) can be written in a different way: $p(x, y) = p(x|y)p(y)$. Similarly, when treating X as the conditional variable, we have $p(x, y) = p(y|x)p(x)$. This means

$$p(x, y) = p(x|y)p(y) = p(y|x)p(x). \tag{4.16}$$

From Equation (4.16), we can write

$$p(x|y) = \frac{p(x)p(y|x)}{p(y)}. \tag{4.17}$$

This is the *Bayes theorem*, which relates the conditional probability of a random variable X given Y (i.e., $p(x|y)$) to the "inverse" conditional probability of variable Y given X (i.e., $p(y|x)$). In the Bayes theorem, $p(x|y)$ is usually called the *posterior probability*; $p(x)$ is called the *prior probability*, $p(y|x)$ is called the *likelihood probability*, and $p(y)$ is the marginal probability for Y. The Bayes theorem plays a key role in sequential Bayesian-filtering-based data assimilation. A more in-depth discussion of the Bayes theorem will be provided in Chapter 6.

4.3 Expectation, Variance, and Covariance

The PDF $p(x)$ of a random variable contains a great amount of information. In many cases, instead of working with the full density, it is more convenient to use the statistical moments of the density. The most commonly used statistical moments are expectation, variance, and covariance.

The *expectation* of a random variable represents the average value one "expects" from all the possible outcomes of the random variable. The expectation is also called the *expected value* or the *mean*. It is defined as

$$E[X] = \int xp(x)dx, \qquad (4.18)$$

where $E[X]$ stands for the expectation of random variable X. One can see that the expectation is the average of all outcomes weighted by the probability density $p(x)$.

For a random variable X with PDF $p(x)$, the expectation of a function of X (e.g., $g(X)$) is computed as

$$E[g(X)] = \int g(x)p(x)dx. \qquad (4.19)$$

Expectation is linear. In particular, when a and b are constants, then

$$E[aX + b] = aE[X] + b. \qquad (4.20)$$

Another quantity of interest is the *variance* of a random variable X, denoted by $\text{Var}(X)$, which is defined by

$$\text{Var}(X) = E[(X - E[X])^2]. \qquad (4.21)$$

The variance provides a measure of the dispersion of X around its mean. It is the expected value of the square of the deviation of X from its mean. It can be shown that

$$\text{Var}(X) = E[(X - E[X])^2] = E[X^2] - E[X]^2. \qquad (4.22)$$

Since the variance has a unit that is the square of the data unit, it is common to use the square root of the variance, which is called the *standard deviation*.

For the normal distribution whose density $p(x)$ is defined by Equation (4.8), it can be shown that its expectation is μ and variance is σ^2 (i.e., the standard deviation is σ).

Different from the variance that measures the variability of a single random variable, the *covariance* is a measure of the joint variability of two random variables. The covariance of any two random variables X and Y, denoted by $\text{Cov}(X, Y)$, is defined by

$$\text{Cov}(X, Y) = E[(X - E[X])(Y - E[Y])] = E[XY] - E[X]E[Y]. \quad (4.23)$$

In general, a positive value of $\text{Cov}(X, Y)$ is an indication that the two random variables tend to increase or decrease at the same time (i.e., the two variables change in the same direction), whereas a negative covariance indicates that the two variables change in opposite directions. When two random variables are independent, then in Equation (4.23), $E[XY] = E[X]E[Y]$ and the covariance $\text{Cov}(X, Y) = 0$.

4.4 Multivariate Distribution

The joint distribution defined in Equation (4.9) can be generalized to multivariate random vectors. An n-dimensional random vector is a vector that includes n random variables. Consider the example of measuring some health characteristics, such as weight, temperature, and blood pressure, of a population of people. Each person's measure is a three-dimensional random vector that includes three random variables: *weight*, *temperature*, and *blood pressure*.

In general, let $\boldsymbol{X} = (X_1, \ldots, X_n)$ be an n-dimensional multivariate random vector (note that we use the boldface letter $\boldsymbol{X}$ to denote a multivariate random vector). The joint PDF of $\boldsymbol{X}$ is the function defined by

$$p(x_1, \ldots, x_n) - p(X_1 = x_1, \ldots, X_n = x_n). \quad (4.24)$$

A commonly used multivariate distribution is the multivariate normal distribution, the PDF of which is defined by

$$p(\boldsymbol{x}) = p(x_1, \ldots, x_n) = \frac{exp(-\frac{1}{2}(\boldsymbol{x} - \boldsymbol{\mu})^T \Sigma^{-1}(\boldsymbol{x} - \boldsymbol{\mu}))}{\sqrt{(2\pi)^n |\Sigma|}}, \quad (4.25)$$

where x is the n-dimensional vector, μ is the n-dimensional mean vector, and Σ is the $n \times n$ covariance matrix. The superscript ^{T} marks the transpose of a vector, and the $|\Sigma|$ stands for the determinant of a matrix. A multivariate Gaussian distribution with mean vector μ and covariance matrix Σ is often denoted by $N(\mu, \Sigma)$.

Let $X = (X_1, \ldots, X_n)$ be a random vector having PDF $p(x_1, \ldots, x_n)$ and $g(x_1, \ldots, x_n)$ be a real-valued function defined on the sample space of X. Then, $g(X)$ is a random variable and the expectation of $g(X)$ is

$$E[g(X)] = \int \cdots \int g(x_1, \ldots, x_n) p(x_1, \ldots, x_n) dx_1 \ldots dx_n. \quad (4.26)$$

Given the joint PDF $p(x_1, \ldots, x_n)$, the marginal distribution of any subset of the coordinates of $(X_1, \ldots, X_n)$ can be computed by integrating the joint distribution over all possible values of the other coordinates. For example, the marginal distribution of the first k coordinates $(X_1, \ldots, X_k)$ is computed as follows:

$$p(x_1, \ldots, x_k) = \int \cdots \int p(x_1, \ldots, x_n) dx_{k+1} \ldots dx_n. \quad (4.27)$$

The conditional distribution of a subset of the coordinates of $(X_1, \ldots, X_n)$, given the values of the remaining coordinates, is obtained by dividing the joint PDF by the marginal distribution of the remaining coordinates. For example, assuming $p(x_1, \ldots, x_k) > 0$, then the conditional distribution of $(X_{k+1}, \ldots, X_n)$, given $X_1 = x_1, \ldots, X_k = x_k$, is the function of $(x_{k+1}, \ldots, x_n)$ defined by

$$p(x_{k+1}, \ldots, x_n | x_1, \ldots, x_k) = \frac{p(x_1, \ldots, x_n)}{p(x_1, \ldots, x_k)}. \quad (4.28)$$

4.5 Monte Carlo Method

Consider the example of computing the expectation of a function g defined on the sample space of a random vector X. Equation (4.26) shows how $E[g(X)]$ can be computed. In many situations, however, it is not analytically possible to compute the above multiple integral exactly. In these situations, one may approximate $E[g(X)]$ using the approach of the *Monte Carlo method*.

The Monte Carlo method is a computational approach that relies on repeated random sampling to obtain numerical results. It is also called the *Monte Carlo experiment* or the *Monte Carlo simulation*. Note that the term "simulation" in the Monte Carlo simulation focuses on the fact that it uses repeated random sampling to obtain the statistical properties of some phenomenon. This is different from the simulation approach described in this book, which focuses on the fact that a simulation model is an abstracted representation of a real system.

Applying the Monte Carlo method to the above example of computing $E[g(\boldsymbol{X})]$, we would generate a random vector instance (also called a *sample*) $\boldsymbol{X}^{(1)} = (X_1^{(1)}, \ldots, X_n^{(1)})$ based on the joint density $p(x_1, \ldots, x_n)$ and compute $Y^{(1)} = g(\boldsymbol{X}^{(1)})$. We then generate a second sample $\boldsymbol{X}^{(2)}$, which is independent of the first one, and compute $Y^{(2)} = g(\boldsymbol{X}^{(2)})$. We keep doing this for k times to have k number of independent values $Y^{(i)} = g(\boldsymbol{X}^{(i)})$, $i = 1, \ldots, k$ for the random variable Y. When k is large enough, by the strong law of large numbers, we have

$$\lim_{k \to \infty} \frac{Y^{(1)} + \cdots + Y^{(k)}}{k} = E[Y^{(i)}] = E[g(\boldsymbol{X})]. \qquad (4.29)$$

In this way, the expectation $E[g(\boldsymbol{X})]$ is approximated by the average of the computed samples of Y. This example shows how the Monte Carlo method can be used to approximate the expectation (or other statistical moments) of a random variable.

Another usage of the Monte Carlo method is to generate samples to approximate the probability distribution of a random variable. Let $x^{(i)}$, $i = 1, \ldots, k$, be the set of samples drawn from a target probability distribution, then the PDF of the distribution can be approximated as

$$p(x) \approx \frac{1}{k} \sum_{i=1}^{k} \delta(x - x^{(i)}), \qquad (4.30)$$

where $\delta(\ )$ is the Dirac delta function.

To illustrate this usage, we consider the example of using the Monte Carlo method to approximate the probability density of the one-dimensional standard normal distribution. A standard normal

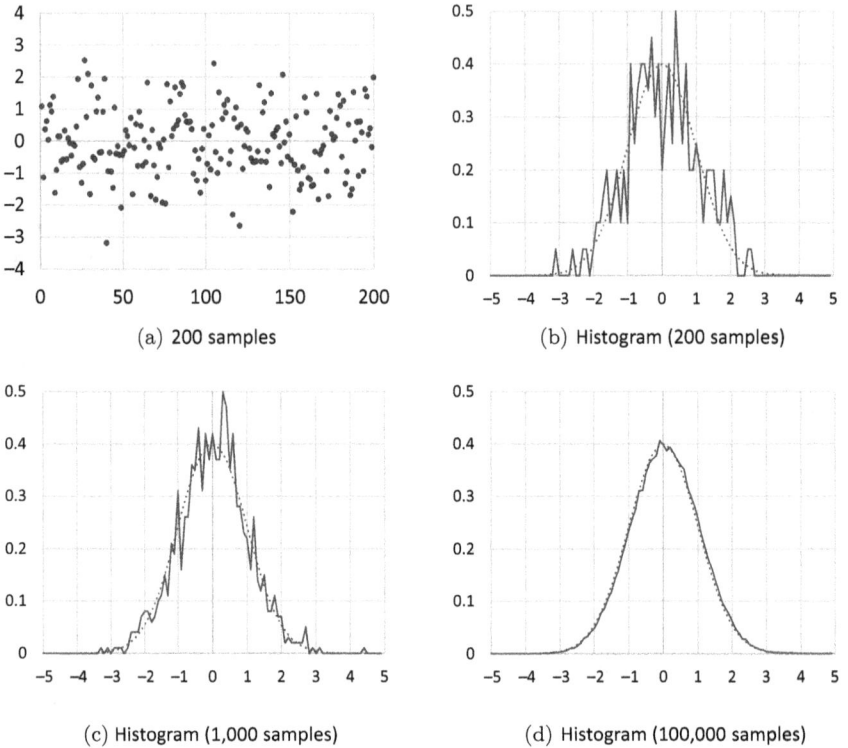

(a) 200 samples

(b) Histogram (200 samples)

(c) Histogram (1,000 samples)

(d) Histogram (100,000 samples)

Figure 4.1.   Illustration of the use of the Monte Carlo method to approximate the standard normal distribution: (a) 200 samples generated from the normal distribution; (b) the normal distribution is approximated by the histogram of the 200 samples shown in (a); (c) approximation result by using 1,000 samples; (d) approximation result by using 100,000 samples. In (b)–(d), the dotted curve is the density curve of the standard normal distribution that needs to be approximated, and the continuous curve is the approximated density based on the histogram.

distribution has $\mu = 0$, and $\sigma = 1.0$, whose PDF is defined in Equation (4.8). Figure 4.1 shows the approximation results. As can be seen, the Monte Carlo method is able to approximate the PDF of the normal distribution when the number of samples is large. The more the number of samples, the more accurate the approximation result. In this example, when the number of samples reaches 100,000, the approximation result becomes very close to the real density.

The two examples described above assume that we are able to generate samples directly from the target PDF. Often, generating samples directly from the target PDF is difficult or impossible.

To solve this problem, different sampling methods have been developed, such as the inverse transform sampling, acceptance rejection sampling, importance sampling, and Markov chain Monte Carlo sampling. The principles of the importance sampling will be described in Section 6.8.1. Details of other sampling methods can be found in the literature and are omitted here.

4.6 Sources

The descriptions in Sections 4.1–4.3 are influenced by Thrun *et al.* (2005) and Evensen (2009). The description of the multivariate distribution (Section 4.4) is based on the work of Casella and Berger (1990). The introduction of the Monte Carlo method (Section 4.5) is influenced by Ross (1997).

Part 2

Dynamic Data-Driven Simulation

Chapter 5

Dynamic Data-Driven Simulation

5.1 What is Dynamic Data-Driven Simulation?

Many simulation applications involve running simulations concurrently with a real system while the system is in operation. For example, a road traffic simulation may simulate a city's real-time traffic; a wildfire simulation may simulate a wildfire that is spreading; a manufacturing system simulation may simulate the operation of an ongoing manufacturing process. These simulations run in a real-time context to provide real-time prediction and analysis for the systems' behavior. The ultimate goal is to support real-time operation or decision-making for the systems under study.

A recurring theme of these simulation applications is to increase the accuracy of simulation results. To this end, much effort has been devoted to improving the fidelity of simulation models. A simulation model embodies a modeler's understanding of a system's behavior as well as the mechanisms that enable it. In general, with more understanding of how a system works, higher fidelity simulation models can be developed. A higher fidelity model captures the behavior of a system in a more faithful way and thus can generate more accurate simulation results.

While a simulation model's quality is important, one should recognize that, regardless of the model used, it is inevitable for a simulation to deviate from the actual process of a real system. This is due to the following errors that commonly exist in simulation applications:

- **Model error:** A model is always different from a real system. The difference can be attributed to the following reasons: (1) A model may be developed based on incomplete knowledge about how a real system works. This means important dynamics or relationships of the real system may not be captured in the model. (2) Even if there is no knowledge gap, i.e., one knows exactly how a real system works, a simulation model is an abstraction of the real system. The abstract representation means that some aspects of the system are omitted or simplified, which leads to differences in the simulation result that can accumulate over time. (3) A system cannot be isolated from its environment, which may shift over time and influence how the real system works. Such changes in the environment may not be adequately modeled by the simulation model. (4) Many dynamic systems are stochastic in nature due to process noises or disturbances that cannot be precisely modeled. The stochasticity in system behavior means that a simulation model cannot produce exactly the same results as those from a real system.
- **Initial state error:** All simulations start from some initial states. In order for a simulation to generate simulation results close to the real system results, it is imperative for a simulation to start from an initial state that is the same as or similar to the real system's state. Unfortunately, in many cases, the true system state is unknown and can only be estimated. The estimated states introduce errors to the initial states, which lead to differences in simulated behavior.
- **Data error:** Complex simulation models often use external data. These include system characteristic data that parameterize simulation models (such as the GIS and fuel data in a wildfire spread simulation) and input data that act as simulation inputs (such as weather inputs in a wildfire spread simulation). Errors in these data make a simulation model behave differently from the real system. Unfortunately, data errors are universal due to the following two fundamental reasons: (1) inaccurate measurements stemming from the data acquisition processes or sensors; (2) limited resolution. The limited resolution may include limited spatial resolution or limited temporal resolution. An example of the former is GIS data that have a spatial resolution of 30 m, meaning that there is only one data point for every 30×30 m region. An example of the latter is weather data that are obtained every 5 min, meaning that there is no weather update during the 5 min time interval.

The limited resolution gives rise to discrete data points that differ from reality, where information changes continuously in space and time.

Due to the above errors, it is common for a simulation to generate results that are different from a real system's results. This difference can grow indefinitely over time, leading to large errors that eventually make the simulation results not useful. Since simulation model is only one of the factors that influence the result of a simulation, hereinafter, we use the term *simulation system* to refer to the system of a simulation. A simulation system includes the simulation model as well as other components, such as the components for determining initial state and external input, that are necessary to set up and run simulations.

One way to address the discrepancy between a simulation system and a real system is to utilize real-time data collected from the real system to dynamically adjust the simulation system. The real-time data carry information about the real-time condition of the real system, which can be utilized to produce more accurate simulation results. As the real system evolves over time, continuous adjustments are needed. The continuous adjustments align the simulation system with the real system on a regular basis to keep the discrepancy between the two bounded. They also provide the opportunity to continuously calibrate the simulation model based on real-time data.

Figure 5.1 illustrates the approach where real-time data are collected and fed into a simulation system regularly. In the figure, t_1, $t_2, \ldots, t_k$ are different time instants, called *time steps*, when real-time data are collected from a real system in operation. At each time step, the simulation system uses the collected data to adjust itself and to run simulations to provide real-time prediction/analysis for the real system. When new data arrive at the next time step, the simulation system is readjusted and new simulations are run to provide new simulation-based predictions/analyses. In this way, the simulation system continuously updates its simulations to provide more accurate predictions/analyses for the system in operation.

The continuous incorporation of real-time data into a simulation system gives rise to a new paradigm of *dynamic data-driven simulation* as defined in the following.

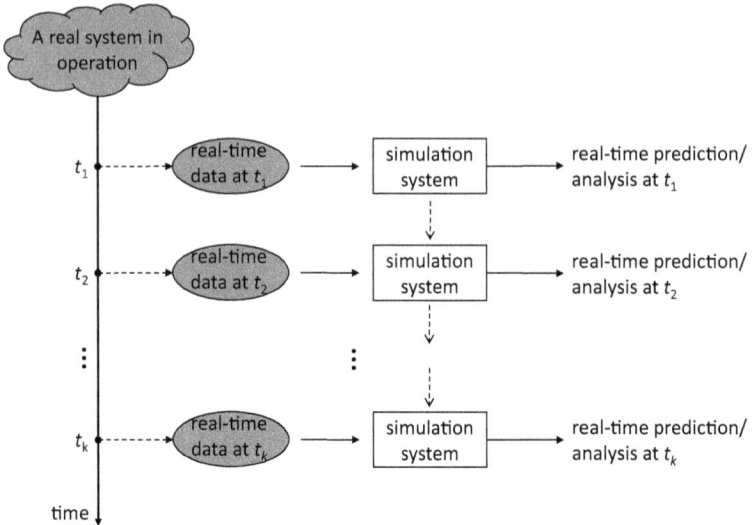

Figure 5.1.   Illustration of dynamic data-driven simulation.

> **Definition:** Dynamic data-driven simulation (DDDS) refers to a simulation paradigm where a simulation system continuously and systematically assimilates real-time data from a system in operation to support real-time prediction and analysis for the system.

Several things are worth mentioning in this definition. First, DDDS runs concurrently with a *system in operation*. The term "system in operation" is used here to emphasize that the system is an actual running system, as opposed to an imaginary system or a system under development. DDDS works with a real system and provides real-time prediction and analysis for the system. The real-time prediction/analysis can then be used to support real-time decision-making or operation for the system. For example, DDDS of a running manufacturing system may predict the system's potential bottlenecks to support real-time job scheduling for the manufacturing system; DDDS of a spreading wildfire may predict the fire's future spread to support decision-making for fire suppression; and DDDS of an ongoing infectious disease spread may predict/analyze the results of potential interventions to support vaccination strategies and medical resource allocations.

Second, DDDS deals with simulation activities. It differs from the modeling activities of a simulation project. DDDS belongs to the simulation phase — it assumes that the underlying simulation model has been developed (e.g., having gone through calibration and validation). DDDS may still adjust the parameters of a model using real-time data, but the adjustment is based on an existing model that has been developed.

Third, DDDS emphasizes continuous and systematic incorporation of real-time data. It differs from other approaches that use real-time data in *ad hoc* ways or treat each use of real-time data in isolation. In DDDS, the incorporation of real-time data is not a one-time action. Instead, it is an ongoing process, where data are assimilated sequentially and regularly. This process view of assimilating real-time data means that the simulation systems at different time steps are not isolated from each other — they are results of an ongoing adjustment process. Furthermore, DDDS emphasizes activities that systematically cover the different aspects of a simulation system. The systematic treatment of DDDS activities (described in the following section) leads to a more organized usage of real-time data for improving simulation results.

Figure 5.2 compares DDDS with offline simulation and conventional online simulation from the real-time data usage point of view. Offline simulation does not use real-time data from a real system. In many cases, the real system does not exist yet (e.g., it is still being designed), as denoted by the dashed box in Figure 5.2(a). Conventional online simulation runs concurrently with a real system and may use real-time data from the real system. However, it does not emphasize continuous and systematic incorporation of real-time data, as it treats each use of real-time data in isolation. This is illustrated by the single arrow between the real system and the simulation system in Figure 5.2(b). On the other hand, DDDS runs concurrently with a real system and is dynamically driven by real-time data. It emphasizes continuous and systematic assimilation of real-time data, as illustrated by the multiple arrows in Figure 5.2(c).

The differences between the three types of simulation can be further explained using the wildfire simulation example. Before a wildfire season starts, a fire manager may run simulations to study fire spread behavior under various weather conditions. This is the case of offline simulation, which does not use any real-time data from a

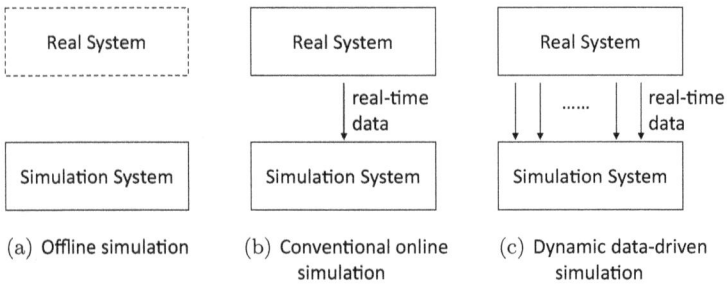

Figure 5.2. Comparing DDDS with offline simulation and conventional online simulation: (a) Offline simulation does not use real-time data; (b) conventional online simulation treats each use of real-time data in isolation; (c) DDDS uses real-time data continuously and systematically.

real wildfire. When a real wildfire breaks out, the fire manager may want to run simulations to study the spread of the fire. These simulations may use real-time data. Nevertheless, the use of real-time data is not systematic, and each use of real-time data is treated in isolation. This is the case of conventional online simulation. In DDDS, the simulation system is live all the time — it assimilates real-time data regularly and systematically to adjust the simulation model. The regular and systematic use of real-time data makes the simulation system always aligned with the dynamically spreading wildfire to enable real-time prediction/analysis.

5.2 DDDS Activities

Given a simulation model, the goal of DDDS is to support accurate simulation-based prediction/analysis for a dynamic system in operation. Several issues need to be addressed in order to achieve this goal.

First, a simulation needs to start from an initial state that matches the actual state of a dynamic system. Since a dynamic system constantly changes its state, a simulation-based prediction should initialize itself with the real-time system state at the time of the simulation run. While this idea is straightforward to understand, it is not always easy to implement. This is because, for many dynamic systems, the real-time state (or some portion of the real-time state) is unknown and cannot be directly derived from the observation data collected from the dynamic systems. To give an example, a wildfire spread simulation needs to start from a fire front describing the initial shape of the fire. Nevertheless, the observation data (e.g., images collected by a drone) at any moment may only cover a (small) portion of the fire perimeter due to the large size of the fire. This raises the issue of *dynamic state estimation*, in which the dynamically changing state of a system is estimated based on real-time observation data so that a simulation run can be correctly initialized.

Second, to predict how a dynamic system works, a simulation system needs to calibrate its model parameters to accurately reflect the real-time characteristics of the system. Such calibration often needs to be done in an online fashion due to the following two reasons: (1) It is common for some characteristics of a system to be known only after the system operates in the real field. This means one cannot simply assign some "typical" or "average" values to the corresponding model parameters. Instead, the parameters need to be estimated in real time based on how the system actually works. (2) Complex dynamic systems may dynamically shift their characteristics due to changes in the operating environment or working conditions. For these systems, the corresponding model parameters are not static and need to be dynamically estimated based on real-time data from the system. This raises the issue of *online model calibration*, in which the parameters of a simulation model are dynamically calibrated based on real-time data from a dynamic system.

Third, many dynamic systems are influenced by external inputs generated from other systems or from the environment. Simulation-based prediction/analysis of a system's future behavior thus needs to take into account the future inputs the system will receive. For example, a manufacturing system's operation is influenced by the incoming jobs that need to be processed. To predict/analyze the manufacturing system's future behavior, it is important to take into consideration the future jobs that will arrive and need to be processed. This raises the issue of *external input modeling & forecasting*, in which a system's future inputs are modeled or forecasted based on its past input data.

Finally, to enable simulation-based prediction/analysis, DDDS needs to set up and run simulations using the right information. This is the task of *simulation-based prediction/analysis*, which integrates the results from other DDDS activities to carry out

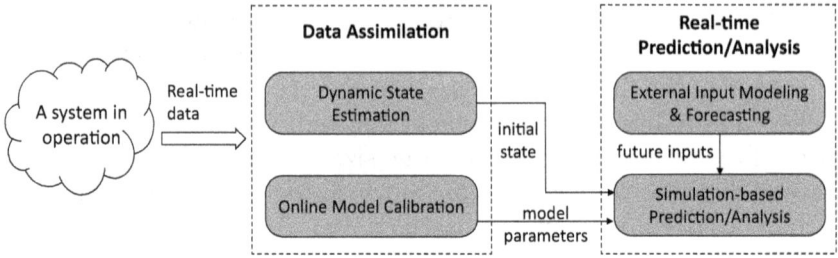

Figure 5.3.   DDDS activities.

simulation-based prediction/analysis. Specifically, simulation-based prediction/analysis should start from the initial states estimated from dynamic state estimation, employ the calibrated simulation model resulting from the online model calibration, and use input trajectories modeled/forecasted by the external input modeling & forecasting.

Each of the above four issues is handled by a corresponding DDDS activity. Figure 5.3 shows a block diagram of the four DDDS activities: dynamic state estimation, online model calibration, external input modeling & forecasting, and simulation-based prediction/analysis.

These activities are placed into two groups based on their functionalities: the *data assimilation* group and the *real-time prediction/analysis* group. Dynamic state estimation and online model calibration belong to the data assimilation group because they deal with "estimation" of the model state or model parameters based on real-time observation data. These estimation tasks can be framed as data assimilation problems and addressed by data assimilation methods (details come later). The two estimation-related activities provide information for aligning a simulation system with the real system at the *current time*. The real-time prediction/analysis group includes external input modeling & forecasting and simulation-based prediction/analysis because both deal with "prediction" of the future. The former models/forecasts the future inputs, and the latter simulates the future behavior. The four activities work in concert. In particular, the simulation-based prediction/analysis uses results from the other three activities, as shown in the figure.

A brief description of the data and the DDDS activities is given in the following. Each activity will be elaborated upon in later sections:

- **Real-time data:** These are the data collected from a real system, reflecting the real-time condition of the system in operation. Real-time data include *real-time measurement data* and *real-time input data*, which are used by data assimilation and input modeling/forecasting, respectively.
- **Dynamic state estimation:** This activity assimilates real-time measurement data to estimate the "current" state of a real system. The estimated state will be used to initialize simulation runs for predicting/analyzing the system's future behavior.
- **Online model calibration:** This activity calibrates the simulation model's parameters based on real-time measurement data to make the model better represent the system in operation. The calibrated model will be used as the simulation model for predicting/analyzing the system's future behavior.
- **External input modeling & forecasting:** This activity models/forecasts the future inputs of a real system based on real-time and historical input data. The forecasted inputs will be used as input trajectories for the simulation runs for predicting/analyzing the system's future behavior.
- **Simulation-based prediction/analysis:** This activity uses simulations to carry out real-time prediction/analysis for a system in operation. It generates prediction/analysis results that can be used to support decision-making or operation for the system.

By definition, DDDS is carried out in an iterative way. In each iteration, real-time data are assimilated to estimate the current system state and to calibrate the simulation model parameters. The real-time data also enable modeling & forecasting of future inputs. The estimated state, calibrated model parameters, and forecasted inputs are then used to set up simulation runs to enable simulation-based prediction/analysis for the system in operation. The next iteration starts when a new batch of real-time data become available. The new iteration incorporates the new data to provide new simulation-based predictions/analyses.

Depending on different applications, some DDDS activities may have more (or fewer) roles than others. For example, if a system does not expect any external input, then the external input modeling & forecasting activity is not needed. Similarly, if a system's characteristics are already well captured by the simulation model at hand, then the online model calibration activity may be skipped.

5.3   Real-Time Data

DDDS runs in parallel with a real system. To align the simulation system with the real system, an information bridge between the two needs to be established. Such an information bridge is served by the real-time data collected from the real system.

Based on their roles in a real system, we differentiate two types of real-time data in DDDS: *input data* and *measurement data*, as defined in the following:

- **Input data** describe a real system's external inputs, which originate from outside of the system and are not under the control of the system itself. Input data contain information about how the real system is influenced by its external environment and thus need to be taken into account for aligning the simulation system with the real system.
- **Measurement data** (also called *measurements*, or *observational data*, or simply *observations*) are data reflecting a system's internal state or output, which are results of the system's dynamics. These data are related to the *observability* of a system, which is the ability to infer a system's internal state by examining its external measurements. Note that in data assimilation, the term "measurement data" specifically refers to the data that reflect a system's internal state or output (the "output" is often omitted because output is a function of internal state).

Distinguishing input data from measurement data is necessary because the two have different relationships with a system and thus play different roles in DDDS.

Input data originate from outside a system. Real-time input data serve two roles in making simulation-based prediction/analysis more accurate. First, it allows a simulation model to reproduce what has influenced the real system so that the simulation model can evolve to the same or a similar state as that of the real system. Corresponding to this role, the input data are used by the dynamic state estimation and online model calibration activities to run the simulation model forward as part of the data assimilation procedure. Second, input data provide information to model or forecast the future inputs that are likely to happen to the real system, which are needed to simulate the system's future behavior. Corresponding to this role, the input data are used by the external input modeling & forecasting activity to model/forecast future inputs.

Measurement data provide information about a system's real-time state. These data are needed while estimating the system's dynamically changing state for initializing simulation runs. They also carry information for calibrating a simulation model's parameters to make the simulation model more accurately characterize the system in operation. Measurement data are used by the two DDDS activities in the data assimilation group to align the simulation system with the real-time condition of the real system. A major part of this book is devoted to describing how to assimilate measurement data in systematic ways using data assimilation methods (Chapter 6).

Measurement data are collected by sensors. The quality of these data depends on the capabilities and characteristics of the sensors. The two most important metrics that characterize a sensor's measurements are *accuracy* and *precision*. Understanding them will help in understanding the uncertainties involved in sensor measurement. Note that several other metrics, such as *resolution* and *sensitivity*, also exist that characterize a sensor's measurement. Readers can refer to the literature to learn more about them.

- Accuracy is a measure of the *degree of closeness* of a measured value to the true value. Accuracy is directly related to measurement error, which is the difference between the measured value and the true value. A higher accuracy means a smaller error, and a lower accuracy means a larger error. For example, a reading of 95° from a high-accuracy temperature sensor would indicate that the true temperature is very close to 95°.
- Precision is a measure of the *reproducibility* of a set of measurements. A sensor with high precision would provide measurements that are highly consistent with each other. In the above

temperature sensor example, it means the different measurements from the temperature sensor would be close to each other (assuming the true temperature does not change). Note that a sensor with high precision may not have high accuracy. For example, a high-precision sensor that is not calibrated properly would consistently return measurements that have similar large errors from the true value.

Figure 5.4 illustrates a sensor's accuracy and precision using the probability density of measured values in comparison to the true value under measurement. In the figure, the measured values are assumed to follow a normal distribution. The accuracy is indicated as the difference between the true value and the mean of the measured values. The precision is indicated as the standard deviation of the measured values. In general, a higher precision would mean a narrower distribution for the different measurements, and a higher accuracy would mean a closer distance between the mean and the true value.

Measurement error can be divided into *random error* and *systematic error*. Random error (also called *noise*) is related to the precision of the sensor. It is an inherent part of any sensor measurement. When the true value under measurement is static, one can reduce the impact of random error by averaging multiple measurements. Systematic error (also called *bias*) is consistent and repeatable error associated with a faulty or uncalibrated sensor or a flawed experiment design. Systematic errors cannot be reduced through averaging. If known, they may be subtracted from the measurement values.

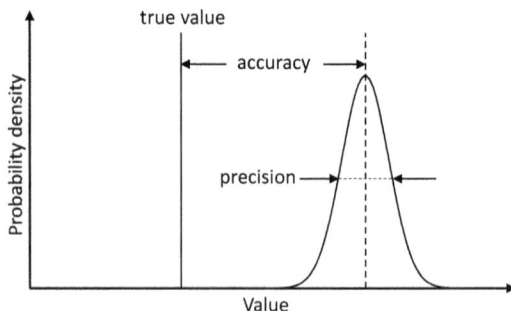

Figure 5.4.　Accuracy and precision of measurement data.

The above discussion shows that measurement data from sensors are not perfect and may involve a great deal of uncertainties. These uncertainties mean that measurement data should not be simply used as the ground truth for a specific value under measurement. Dealing with noisy measurement data is one of the key considerations when assimilating real-time measurement data into simulation models.

5.4 Dynamic State Estimation

5.4.1 *The state estimation problem*

The dynamic state estimation activity estimates the dynamically changing state of a real system based on real-time measurement data from the system. It addresses the simulation initialization problem by providing initial states for simulation-based predictions/analyses. The problem of state estimation is fundamental to DDDS because simulation runs need to be initialized based on the real-time state of a system in operation. This is different from offline simulations, where initial states are set based on prior knowledge or hypothetical scenarios.

There is a misconception that state estimation is a trivial task because one can directly obtain the state from measurement data. While this may work for simple applications, in most cases, the initial state of an online simulation should not and cannot be directly set based on measurement data, as explained in the following.

First, measurement data are noisy and have errors. Setting states directly using noisy data can lead to initial states that differ significantly from the real system states. More importantly, it can result in infeasible or erroneous states for a simulation model. Figure 5.5 illustrates this issue using a scenario of a mobile robot moving in a hallway corridor. In the figure, the horizontal open space in the middle is the corridor, the other open spaces are office rooms, and the gray areas are walls. The robot is equipped with a GPS sensor that reports its position regularly (e.g., every second). A controller uses an online simulation to help plan the robot's movement — it runs a simulation in every step to predict/analyze the robot's position in the near future. In this example, since the GPS sensor provides the robot's real-time position at each step, one would be tempted to set the initial state of a simulation directly based on the GPS data at

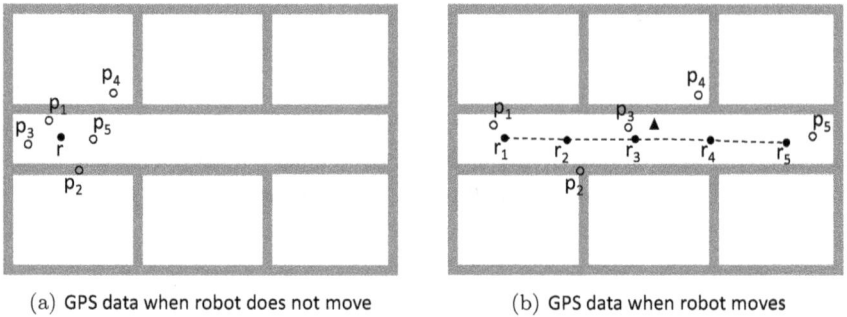

(a) GPS data when robot does not move   (b) GPS data when robot moves

Figure 5.5. Illustration of the limitations of measurement data.

the time of a simulation run. However, it is known that GPS sensors have errors (for example, a typical GPS sensor on a smartphone can have errors of up to 4.9 m under an open sky, and its accuracy worsens near buildings, bridges, and trees (GPS.gov, n.d.)). This means a real-time GPS measurement would be different from the true position of the robot.

Figure 5.5(a) illustrates a scenario when the robot does not move. The robot's true position is displayed as a solid circle, and the real-time GPS data at five consecutive time steps are displayed as empty circles (denoted by p_1, p_2, ..., p_5). If each time step's simulation uses the corresponding GPS data as the initial position of the robot, one can see that p_2 is infeasible because it is within a wall and that p_4 is an erroneous state because it is in a different office room, which will make the simulation very different from how the real robot works. One may try to average the GPS data from several measurements to reduce noise. This averaging approach works only if the robot does not move. When the robot moves, the averaged positions would lag the true positions of the robot, not to mention that they may still result in infeasible or erroneous states. Figure 5.5(b) illustrates a scenario when the robot moves. The true positions of the robot at five consecutive time steps are denoted by r_1, r_2, ..., r_5 (solid circles) and the corresponding GPS data are denoted by p_1, p_2, ..., p_5 (empty circles). The averaged position from p_1, p_2, ..., p_5 is shown as a triangle, which significantly lags behind the true position r_5. This example shows that even when real-time measurement data are directly related to a state of interest, setting the state directly based on the data may not work.

Another reason that makes state estimation necessary is that measurement data are discrete and may be partial or incomplete. Measurement data are generally collected by sensors using sampling techniques that render data at discrete time instants. There is no data available between the time instants. Meanwhile, for many systems, it is not practical to have sensors to cover the entire system without any "gap" in data collection. This is especially true for spatiotemporal systems, such as traffic systems and wildfires whose states span across large areas. For these systems, the deployed sensors typically cover a limited number of locations. For the locations that are not covered by sensors, no information is available. Thus, there is a need to estimate the overall system state based on limited or sparse measurements.

An equally important reason that makes state estimation an indispensable task for some applications is the intractability of the inverse problem of deriving state values from measurement data. Consider the example of a wildfire where measurement data are collected from temperature sensors. The state that needs to be estimated is the fire front locations. It is infeasible to derive the precise fire front locations based on temperature measurements because, in theory, there are infinite number of fire front locations that can lead to the same temperature readings. The intractability of the inverse problem calls for effective methods to optimally estimate the dynamically changing state of a system based on real-time measurement data.

The above discussions reveal the limitations of measurement data. In general, a system's state is "internal" and measurement data are "external" observations reflecting some information of the state. Estimating the dynamically changing "internal" state from "external" observations is an essential problem for DDDS that needs to initialize simulation runs based on real-time measurement data. This problem is made more difficult as measurement data have noise and may be sparse or incomplete. Figure 5.6 illustrates the state estimation problem of estimating a system's internal state from measurement data.

Formally, we formulate the dynamic state estimation problem as follows. Let $t_0, t_1, \ldots, t_k$ (with $t_0 < t_1 < \cdots < t_k$) be the sequence of time instants when measurement data are collected, where $k \in \mathbb{N}^0$ is the time index ($\mathbb{N}^0$ is the set of non-negative integer numbers). These time instants may or may not be equally spaced (i.e., having

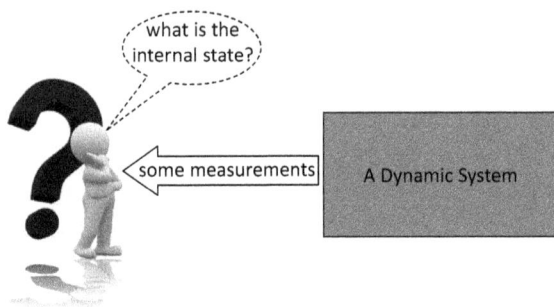

Figure 5.6.   The state estimation problem.

the same time interval). Let x be the state vector and y be the measurement vector. The dimensions of these vectors are application specific. We define $y_k := y(t_k)$ as the measurement y at time t_k and $y_{0:k} := (y(t_0), y(t_1), \ldots, y(t_k))$ as the sequence of measurements up to time t_k. Similarly, we define $x_k := x(t_k)$ as the state x at time t_k and $x_{0:k} := (x(t_0), x(t_1), \ldots, x(t_k))$ as the sequence of states up to time t_k. The task of dynamic state estimation at time t_k is to estimate the state vector x_k based on measurements $y_{0:k}$, i.e., the measurement at time t_k as well as previous measurements. As an estimation problem, this is usually formulated in a probabilistic way. Let $p(x)$ describe the probability distribution of x. The dynamic state estimation is defined as

$$p(x_k|y_{0:k}), \tag{5.1}$$

i.e., computing the probability density of x_k conditioned on the measurements $y_{0:k}$.

The dynamic state estimation is carried out in an iterative way: When a new measurement becomes available, a new state estimation is carried out to estimate the real-time state at that time. For example, when y_{k+1} arrives at time t_{k+1}, a state estimation would be carried out to estimate the state x_{k+1} based on measurements $y_{0:k+1}$.

The above formulation assumes the state estimation is carried out at each time instant when a new measurement is collected. For applications where measurement data are collected in high frequency, estimating state whenever a new measurement arrives may not be necessary or practical. For these applications, it is common to carry out state estimation based on fixed time steps. Let t_0, t_1, ..., t_k be the time steps when state estimations are carried out. We define y_k to include all the measurement data between t_{k-1} and t_k. Then, the problem can be formulated in the same way as (5.1), i.e., estimating the state at t_k based on measurements in the most recent time interval and the pervious time intervals.

5.4.2 *Data assimilation for state estimation*

The formulation in Equation (5.1) does not prescribe how state estimation is carried out. Previous discussion has shown that deriving state values directly from measurement data is not desirable and, in many cases, infeasible. To achieve accurate state estimation, it is necessary to combine additional information with measurement data. Conceptually, it is desirable to use information about the behavior of a system to address the limitations and information gaps of measurement data. Such information can be provided by a dynamic model (i.e., a simulation model) of how the system behaves. Since the simulation model embodies our understanding of how the dynamic system works, it provides us a way to objectively and meaningfully generate trajectories of state change to fill the information gaps or to cross-check the information from measurement data. For example, in the scenario shown in Figure 5.5(b), a simulation model would reveal that p_2 and p_4 are infeasible because p_2 is within a wall and p_4 is in a room that is unreachable from the previous state.

This approach of combining a dynamic model with measurement data for state estimation is referred to as *data assimilation*. Since both model and data have errors and limitations, data assimilation aims to make the best use of both information in a systematic way to achieve state estimation and uncertainty quantification.

Figure 5.7 illustrates how data assimilation works to address the dynamic state estimation problem. The dynamically changing state of a real system (i.e., the true state) is represented by the continuous gray curve, which is hidden and needs to be estimated (in this example, x_0 is assumed to be known). The task of data assimilation is to estimate the true states in real time for time steps t_1, t_2, t_3, $\ldots$ based on the noisy measurement data at those time steps as well as an imperfect simulation model.

The data assimilation is carried out in an iterative way. At time t_1, the simulation model is initialized with state x_0, and it runs for

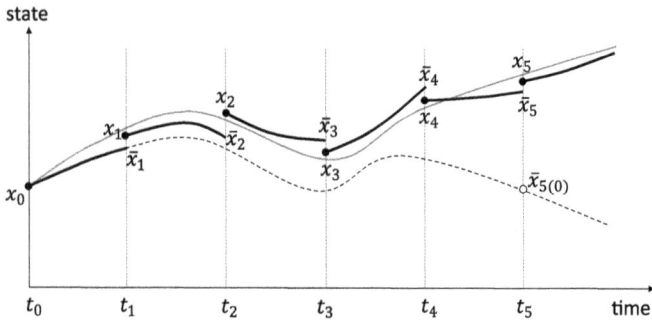

Figure 5.7.   Data assimilation for dynamic state estimation: The true state that is unknown and needs to be estimated is shown by the continuous gray curve. In each step, data assimilation uses the simulation model to predict a new state (denoted by $\bar{x}_k, k = 1, 2, \ldots$), starting from the estimated state from the previous step. It then uses real-time measurement data to adjust the predicted state. The adjusted state (denoted by $x_k, k = 1, 2, \ldots$) becomes the estimated state for the current step. The dashed curve represents a simulation that is not adjusted by data assimilation.

$(t_1 - t_0)$ time to generate a "predicted" state $\bar{x}_1$. The predicted state is then adjusted based on the measurement data (not shown in the figure) at time t_1. The adjusted state (denoted by x_1, displayed as a solid circle) becomes the estimated state at time t_1. In the next iteration, the simulation uses x_1 as the initial state to predict the state at time t_2. The predicted state $(\bar{x}_2)$ is then adjusted using the measurement data at time t_2 to generate a new estimated state (x_2). This process continues for all other data assimilation steps. As a result, when the true state changes, the data assimilation dynamically adjusts the estimated states to follow the true state.

For comparison purpose, a simulation that starts from the same initial state x_0 but does not use data assimilation is shown as the dashed curve in the figure. One can see that the discrepancy between that simulation and the true state grows over time, making the simulation results not useful after several steps. On the other hand, using data assimilation, a simulation-based prediction/analysis can be initialized with the estimated state that is closer to the true state. For example, a simulation-based prediction/analysis at time t_5 would be initialized from x_5, as opposed to $\bar{x}_{5(0)}$ (which represents the simulated state starting from the initial state x_0). This makes the simulation-based prediction/analysis more accurate compared to when no data assimilation is used.

In summary, dynamic sate estimation is a fundamental activity of DDDS and is addressed by data assimilation. Data assimilation systematically "corrects" the estimated state using information from both the simulation model and measurement data. This compares to *ad hoc* approaches of setting initial states based on measurement data. The topic of data assimilation will be elaborated upon in Chapter 6.

5.5   Online Model Calibration

The online model calibration activity dynamically calibrates a simulation model's parameters based on real-time measurement data to make the model more accurately capture the characteristics of a system in operation. Online model calibration is essentially an estimation problem to estimate the values of parameters that need to be calibrated based on real-time measurement data. Let $y_k := y(t_k)$ be the measurement y at time t_k and $y_{0:k} := (y(t_0), y(t_1), \ldots, y(t_k))$ be the sequence of measurements up to time t_k. Let θ be the parameter vector to be calibrated. We define $\theta_k := \theta(t_k)$ as the parameter vector θ at time t_k. Similar to the state estimation problem, the online model calibration problem is defined as

$$p(\theta_k | y_{0:k}), \tag{5.2}$$

i.e., computing the probability distribution of θ_k conditioned on the measurements $y_{0:k}$. Online model calibration is carried out in an iterative way: When new measurement data become available, new calibrations are carried out to update the estimations of the model parameters.

In some applications, one may estimate a parameter's value directly from measurement data. For example, the *production rate* of a manufacturing machine may be a parameter that needs to be calibrated in real time based on how the machine works in the real field. The value of this parameter may be derived directly from the measurement data of job processing time (assuming such measurements are available).

More generally, due to the same limitations of measurement data described in the previous section, it may be undesirable or infeasible to derive parameter values directly from measurement data. For these applications, a common approach is to formulate the online model calibration as a joint state–parameter estimation problem. With this approach, the parameters that need to be calibrated are treated as part of the state vector that needs to be estimated. Then, the same data assimilation method developed for state estimation can be applied to estimate the state and parameter values at the same time. The joint state–parameter estimation problem is described in Chapter 6.

It is worth noting the difference between the online model calibration and the calibration work used in developing a simulation model (called *offline calibration* in the following description). Offline calibration is the process of finding a set of model parameters to improve model fit, i.e., to make the simulation results match historical data or desired outputs. Offline calibration is often formulated as a global optimization problem using historical data that may span a long period of time, e.g., days or years. This compares to online calibration that works in an iterative way, where each iteration uses newly arrived real-time data to continuously update the estimation of the model parameters. Another major difference is that offline calibration deals with a complete set of model parameters, whereas online calibration typically focuses on a subset of parameters that need to be calibrated or tend to vary during a system's operation.

To achieve effective online calibration, one needs to analyze the application and the simulation model to decide the "essential" set of model parameters that need to be calibrated. Consider the example of a freeway traffic simulation. Parameters that may not change as rapidly, such as the influence of road condition on vehicles' speed, may be assumed to be fixed. Long-term changes in the behavioral patterns, such as ratios of cars or trucks, could be reflected through periodic offline recalibration. Only the parameters that can vary significantly in real time, such as the preferred safe distance in car following, which can change quickly due to weather conditions, such as rain, are candidates for online calibration.

Despite their major differences, offline calibration and online calibration are related. In particular, offline calibration provides a baseline model, based on which online calibration is carried out. Figure 5.8 shows this relationship. Offline calibration belongs to the modeling phase and uses historical data, where all the data are available at the beginning of the calibration. Online calibration uses the

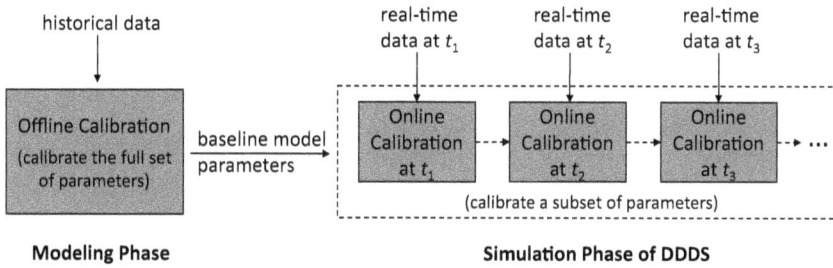

Figure 5.8.   Offline calibration and online calibration.

baseline model parameters resulting from offline calibration. Online calibration belongs to the simulation phase of DDDS and uses real-time data. It is an iterative process involving multiple steps, each of which incorporates new real-time data that become available in the most recent time interval.

5.6   External Input Modeling & Forecasting

The external input modeling & forecasting activity models and forecasts future inputs of a system based on the system's real-time and past input data. This activity is necessary because a simulation-based prediction/analysis needs to take into account the future inputs that influence a system's behavior. External input modeling & forecasting generates input trajectories that are used by a simulation run for predicting/analyzing a system's future behavior.

It is useful to distinguish two types of external input a system may receive: *control input* and *environment input*. Control inputs are external inputs originating from a system operator (or decision maker) to control the system's operation. For example, a manufacturing system may receive a control command to change its operating mode to an "energy-saving mode". The issuing of a control input is at the discretion of the system operator. Environmental inputs are external inputs originating from the environment or other systems connected to the system. These inputs originate from outside the system, which is not at the discretion of the system operator. For example, a manufacturing system may receive jobs that need to be processed from another factory. The generation and arrival of these jobs (e.g., job type, size, and timing) are determined by the external factory, over which the system operator has no direct control.

The two types of inputs can be handled differently when simulating a system's future behavior. Typically, there is no need to forecast the future control inputs because these inputs are at the discretion of the system operator. Instead, future control inputs are either assumed to be known (e.g., provided by the system operator or a decision-making model) or treated as a decision choice to be evaluated in a what-if analysis. For example, to analyze the performance of operating a manufacturing system in an "energy-saving mode" or a "normal-operation mode," the two alternatives of control input would be injected in the simulation runs of different what-if analyses, and their results are compared.

On the other hand, a system's future environment inputs are not under the control of the system operator. Thus, these inputs need to be forecasted or modeled so that their influence on the system can be simulated. For the manufacturing system example described above, one needs to model/forecast the jobs that will arrive in the future and then use the forecasted arrivals of jobs as input trajectories for the simulation runs that predict/analyze the manufacturing system's future behavior.

The rest of this section focuses only on environment input (the qualifier "environment" will be omitted when the context is clear) to discuss external input modeling & forecasting. We note that input modeling & forecasting may not be needed if a system does not expect any input or the input can be obtained from another source. For example, a wildfire spread simulation needs external weather input, which can be obtained from a weather forecast service.

When input modeling/forecasting is necessary, the activity of input modeling & forecasting is used to generate input trajectories to accurately represent a system's future inputs. Often, the generation of input trajectories is based on a model. In the following, we describe three approaches of input modeling & forecasting.

5.6.1 *Modeling with probability distribution*

This approach models the distribution of external inputs using a probability distribution and then generates input trajectories by drawing samples from the probability distribution. This is the traditional way of simulation input modeling. In this approach, a probability distribution is selected and its parameters are fit based on historical input data. The fitted probability distribution then acts as the input model to generate inputs for simulation runs. For example, the number of jobs received in each day by a manufacturing system may be modeled by a Poisson distribution whose rate parameter λ can be determined by fitting with historical job data. To generate jobs in future days for a simulation run, a sequence of samples is drawn from the fitted Poisson distribution to represent the sequence of jobs in future days.

The probability distribution modeling approach is commonly used for generating input trajectories for offline simulations. Nevertheless, standard probability distribution models generally assume independence among samples and thus do not work well in real-time forecasting, where inputs typically have a dependence on time. For example, an observation of high numbers of jobs received by a manufacturing system in recent days is likely to be followed by high numbers of jobs in the near future (e.g., due to being in a busy season). To model this type of time-related dependency, more advanced probability distribution models need to be developed (some examples can be found in the work of Biller and Gunes, 2010).

5.6.2 *Simulation modeling*

This approach develops a simulation model to model how inputs are generated and then uses the simulation model to generate input trajectories. A simulation model differs from other models as it explicitly models the mechanism of input generation. This means some knowledge about the external system that produces the inputs is necessary in order to develop the simulation model. For example, if a manufacturing system receives jobs from a factory, then some knowledge about how the factory produces jobs (e.g., based on an internal schedule) would be needed in order to develop the simulation model.

After an input simulation model is developed, it is coupled with the simulation model of the system under study to form a larger model to support simulation runs. To make the input generation match the prevailing condition of a real system, one may estimate/calibrate the state/parameters of the input simulation model using real-time measurement data in the same way as in dynamic state estimation and online model calibration described earlier. In this way, the external input modeling & forecasting problem essentially becomes a data assimilation problem based on the input simulation model.

The simulation modeling approach has the advantage that it has the potential to model the true mechanism of input generation. Nevertheless, it needs extra knowledge about the external system that produces inputs. In many cases, such knowledge is limited, or it would be too expensive to develop/validate the input simulation model. An alternative approach is to forecast external input purely based on data, as in the *time series analysis* described in the following.

5.6.3   *Time series analysis*

Time series analysis deals with *time series data*, which are data collected sequentially in time. For the problem of input forecasting, the time series data are input data collected over time. Let $t_0, t_1, \ldots, t_k$ be the time steps when input data were collected, $u_k := u(t_k)$ be the input u at t_k and $u_{0:k} := (u(t_0), u(t_1), \ldots, u(t_k))$ be the sequence of inputs up to t_k. Here, u_k can be a scalar quantity, such as the number of jobs arriving at a manufacturing system in a given day, or a k-dimensional vector including k scalar quantities that are recorded simultaneously.

Time series forecasting is the use of a *data model* (different from a simulation model) to predict the future values of a time series based on previously observed values. Given a time series data $u_{0:k}$, let $\mathcal{M}$ be the time series forecasting model, then the one-step-ahead forecasting task at time t_k provides a prediction (denoted by $\vec{u}_{k+1}$) of u_{k+1} based on the past time series data $u_{0:k}$, i.e.,

$$\vec{u}_{k+1} = \mathcal{M}(u_{0:k}). \tag{5.3}$$

This is done in an iterative fashion — when the data u_{k+1} becomes available at time t_{k+1}, the one-step-ahead forecasting then predicts u_{k+2} based on data $u_{0:k+1}$.

The n-step-ahead forecasting task at time t_k predicts a sequence of n future values $\vec{u}_{k+1}, \vec{u}_{k+2}, \ldots, \vec{u}_{k+n}$ based on the past data $u_{0:k}$. There are two basic strategies to achieve multi-step-ahead forecasting: *recursive multi-step forecasting* and *direct multi-step forecasting*. The recursive multi-step forecasting uses a one-step model to predict the future values of a time series in a step-by-step manner.

The prediction of the prior time step is used as an input for making a prediction on the following time step. Given time series data $u_{0:k}$, let $\mathcal{M}$ be the one-step forecasting model, then the n-step-ahead forecasting is carried out in the following way:

$$\vec{u}_{k+1} = \mathcal{M}(u_{0:k}),$$
$$\vec{u}_{k+2} = \mathcal{M}(u_{0:k}, \vec{u}_{k+1}), \qquad\qquad (5.4)$$
$$\vdots$$
$$\vec{u}_{k+n} = \mathcal{M}(u_{0:k}, \vec{u}_{k+1}\vec{u}_{k+2}, \ldots, \vec{u}_{k+n-1}).$$

This recursive multi-step forecasting has the advantage that only one model is needed. Recursive computation is also easy to understand, as each prediction essentially does the same thing: making a one-step forward prediction based on the data before the prediction step. On the other hand, because recursive forecasting uses predicted data as inputs, it allows prediction errors to accumulate, which can make the performance degrade quickly as the prediction time horizon increases.

Direct multi-step forecasting predicts the future values of a time series using separate models. This approach involves developing multiple models — one model for each forecast time step. Let $\mathcal{M}_i$ be the model for the ith-step prediction $(i = 1, 2, \ldots, n)$; the n-step-ahead prediction using the direct multi-step forecasting approach would be carried out in the following way:

$$\vec{u}_{k+1} = \mathcal{M}_i(u_{0:k}),$$
$$\vec{u}_{k+2} = \mathcal{M}_2(u_{0:k}), \qquad\qquad (5.5)$$
$$\vdots$$
$$\vec{u}_{k+n} = \mathcal{M}_n(u_{0:k}).$$

Direct multi-step forecasting uses only actual data for all the predictions and thus does not have the error accumulation issue described above. Nevertheless, because it does not use predictions from previous steps, it means that the dependencies between the predictions cannot be modeled and utilized. Utilizing such dependencies is often important in time series because the prediction on one step can provide information (e.g., indicating a certain trend) for the prediction on the next step.

The two strategies can also work together to support multi-step forecasting. For example, a direct–recursive hybrid method may be developed so that a separate model is constructed for each time step to be predicted, but each model uses the predictions made by models at prior time steps as input values. The predictions are carried out in a step-by-step fashion so that the later steps can use the predictions from the previous steps. Hybrid approaches like this can help to overcome the limitations of the two basic strategies.

The time series forecasting model $\mathcal{M}$ is at the center of any predictions. Various models have been developed, all aimed at improving accuracy by minimizing forecasting errors. The majority of models used in time series forecasting are based on statistical methods, such as smoothing and regression. These models can be referred to as classical time series models. In recent years, there has been an increasing interest in applying machine learning methods, especially those based on artificial neural networks (ANNs), to time series forecasting. The machine learning methods involve using past time series data to train a machine learning model (e.g., a multi-layer ANN) that maps inputs (e.g., time series data in the past 20 steps) to outputs (e.g., time series data in the next 5 steps). The trained model is then used to provide predictions by feeding the model with input data at the time of the predictions.

In the following, we present two basic models as examples of time series forecasting: *simple moving average* (SMA) model and *simple exponential smoothing* (SES) model. There exist many other models, such as autoregressive models and models combining moving average and autoregressive approaches. We refer readers to the literature for more information about these models as well as the machine learning and deep learning models developed in recent years.

5.6.3.1 *Simple moving average*

A moving average model calculates the average of the most recent data points. The assumption behind moving averaging is that a time series is locally stationary with a slowly varying mean. Hence, one can take a local average of the most recent data points to estimate the current mean and use that as the forecast for the near future. The number of data points used in computing the local average is defined by a window size called *window width*. When calculating the average values at different time steps, the window defined by the window width is slid along the time series. A moving average has the effect of smoothing out the fine-grained variation between time steps, and the level of smoothing is influenced by the window width. The larger the window width, the "smoother" the computed values because the averages are calculated from more data points.

An SMA model calculates the average as the unweighted mean of all the most recent data points within the window width. Let $u_{0:k}$ be the sequence of input data up to time t_k. Using the SMA model, the one-step-ahead input forecasting at time t_k would be

$$\vec{u}_{k+1} = \mathcal{M}_{SMA}(u_{0:k}) = \frac{1}{p} \sum_{i=k-p+1}^{k} u_i, \tag{5.6}$$

where $\mathcal{M}_{SMA}(\ )$ represents the SMA model and p is the moving average window width.

5.6.3.2 *Simple exponential smoothing*

The SMA model described above treats the most recent p data points equally and ignores all preceding data points out of the window width. Alternatively, one may consider all the past data but discount them in a gradual fashion — the most recent data point has more weight than the second most recent data point, which has more weight than the third most recent data point, and so on. The exponential smoothing models accomplish this by assigning exponentially decreasing weights to data points going back over time.

The simplest form of exponential smoothing, called SES, uses a smoother factor $\alpha(0 \leq \alpha \leq 1)$ to define how fast the weights decrease over time. Let $u_{0:k}$ be the sequence of inputs up to time t_k. Using the SES model, the one-step-ahead input forecasting at time t_k is defined as

$$\vec{u}_{k+1} = \mathcal{M}_{SES}(u_{0:k}) = \alpha u_k + (1 - \alpha)\vec{u}_k, \tag{5.7}$$

where $\mathcal{M}_{SES}(\ )$ represents the SES model and $\vec{u}_k$ is the prediction of u_k at the previous time step (i.e., at time t_{k-1}).

Since $\vec{u}_k$ was predicted using the same Equation (5.7), we can substitute $\vec{u}_k$ into Equation (5.7) using its defining equation.

Doing this recursively, we have

$$\vec{u}_{k+1} = \alpha u_k + (1 - \alpha)\vec{u}_k$$
$$= \alpha u_k + \alpha(1 - \alpha)u_{k-1} + \alpha(1 - \alpha)^2 \vec{u}_{k-1}$$
$$\vdots$$
$$= \alpha[u_k + (1 - \alpha)u_{k-1} + (1 - \alpha)^3 u_{k-2} + \cdots + (1 - \alpha)^{k-1} u_1]$$
$$+ (1 - \alpha)^k u_0. \tag{5.8}$$

Note that in the above expansion, $\vec{u}_0$ is given the value of u_0. This is generally the case for initializing the first forecast when no prior data is available. Because $(1 - \alpha)$ is between 0 and 1, this expanded equation shows that the weights assigned to the previous inputs decrease exponentially as time goes back.

The smoothing factor α plays a key role in the exponential smoothing model. In general, a larger α would give greater weight to recent changes in the data, while a smaller α would make the prediction less responsive to recent changes. When $\alpha = 1$, $\vec{u}_{k+1} = u_k$, meaning that the predicted next input would be the current input; when $\alpha = 0$, $\vec{u}_{k+1} = \vec{u}_k$, meaning that the current input does not change the prediction because the predicted next input would be the same as the previous prediction.

5.7 Simulation-Based Prediction/Analysis

The ultimate goal of DDDS is to provide real-time prediction and analysis for a system in operation. This goal is served by the simulation-based prediction/analysis activity. The simulation-based prediction/analysis works with other DDDS activities and utilizes information from them: It utilizes the estimated state from the dynamic state estimation activity to initialize simulation runs, the estimated model parameters from the online model calibration activity to parameterize the simulation model for the simulation runs, and the forecasted inputs from the external input modeling & forecasting activity to feed input trajectories to the simulation runs. Because the estimated state, model parameters, and forecasted inputs are updated regularly based on real-time data, together they make the simulation runs aligned with the real system to achieve more accurate simulation results.

Let $\mathcal{F}_{sim}$ be a function representing the simulation operation using a simulation model of interest. At time step t_k, let x_k be the estimated state from dynamic state estimation, θ_k be the estimated model parameters from online model calibration, and $\vec{u}_{k+1:k+i} := (\vec{u}_{k+1}, \ldots, \vec{u}_{k+i})$ be the forecasted input trajectory between t_k and $t_{k+i}(i = 1, 2, \ldots)$ from external input modeling & forecasting. Then, the simulated/predicted system state at time $t_{k+i}(i = 1, 2, \ldots)$, denoted by $\overline{x}_{k+i(k)}$, is a result of the simulation operation in the form of

$$\overline{x}_{k+i(k)} = \mathcal{F}_{sim}(x_k, \theta_k, \vec{u}_{k+1:k+i}, t_{k+i} - t_k), \qquad (5.9)$$

where the term $t_{k+i} - t_k$ is the forward simulation time.

A key feature of simulation-based prediction/analysis is that the simulation runs are not carried out in a single step — they are carried out iteratively to provide continuous updates of simulation-based predictions/analyses. In each step, the simulations use new information that was not available in the previous steps. The new information stems from the new real-time data collected from the real system. Figure 5.9 illustrates the iterative updates of simulation-based predictions/analyses.

Consider a specific step k. We differentiate three state-related concepts:

- *True state* (denoted by x_k^{true}): This is the state of the real system. The true state is not known and needs to be estimated. The true state is not shown in Figure 5.9.
- *Estimated state* (denoted by x_k): This is the estimated state from the dynamic state estimation. Since the true state is unknown, the estimated state is used to initialize simulation runs at each step.
- *Simulated/predicted state* (the simulated state for time t_{k+i} $(i = 1, 2, \ldots)$ is denoted by $\overline{x}_{k+i(k)}$): This is the simulated state generated from a simulation run at step k. Since the simulations at different steps start from different initial states, they generate different state trajectories for the future.

The iterative simulation-based predictions mean that for a specific time in the future, there are multiple predictions from the past

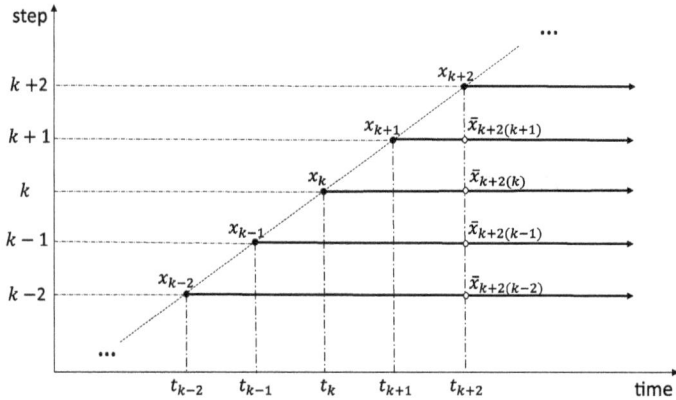

Figure 5.9. Iterative simulation-based predictions in DDDS: The horizontal axis represents time; the vertical axis represents the corresponding steps when simulation-based prediction/analysis is carried out; the thick horizontal lines represent the predicted state trajectories from the simulation runs at different steps. Simulations at each step start from the estimated state for that step (e.g., x_k for step k, displayed as a black circle) to simulate/predict the future.

of that time. Taking time t_{k+2} in Figure 5.9 as an example, a prediction of the state $\overline{x}_{k+2(k)}$ is generated from the simulation run in step k. Before that, each of steps $k-2$ and $k-1$ had generated a prediction for the state (denoted by $\overline{x}_{k+2(k-1)}$ and $\overline{x}_{k+2(k-1)}$, respectively, in Figure 5.9) too. When time reaches t_{k+1}, another prediction $\overline{x}_{k+2(k+1)}$ would be generated. Each of these predictions updates the predictions from the previous steps. In general, the prediction from the most recent step would be the most accurate one because it incorporates new information that was not available in the past predictions.

Several other things are worth mentioning for the simulation-based prediction/analysis activity:

- First, because data assimilation represents the estimated state and model parameters using probability distributions, simulation-based prediction/analysis should employ the Monte Carlo method (see Chapter 4) to run many simulations as opposed to running a single simulation. Each simulation's initial state and model parameters are sampled from their corresponding probability distributions. The simulation results are presented in probability distributions too so that their uncertainty information is preserved.

- Second, simulation-based prediction/analysis may not be invoked in every step of data assimilation. It is invoked only when there is a need, for example, to provide information for decision-making. When simulation-based prediction/analysis is invoked, it always uses the most recent data assimilation results to set up the simulation runs.
- Third, simulation-based prediction/analysis uses a forward time window (from the current time) to define how far to simulate into the future. The size of the time window depends on specific applications. One should be aware that the further one simulates into the future, the less accurate the simulation/prediction results.
- Fourth, the real-time requirement of simulation-based prediction/analysis means that the simulations in DDDS need to run as fast as they can to generate results. For applications with complex simulation models, high-performance simulation is needed in order to meet the real-time requirement of DDDS.

Our discussions so far have treated simulation-based *prediction* and *analysis* together without differentiating the two. Real-time simulation-based prediction and analysis are closely related but have slightly different meanings. Simulation-based prediction predicts a system's behavior based on the most likely path a system will take in the future. Simulation-based analysis, on the other hand, analyzes how the system will behave under various hypothetical scenarios that are intentionally introduced for testing/evaluating the system under study. For example, a what-if analysis may introduce a specific change to the system and then simulates its impact. Despite this difference, both simulation-based prediction and analysis in DDDS must pertain to the real process of the system under study and thus need to be aligned with the real-time condition of the system. A simulation-based analysis can be viewed as a simulation-based prediction under a specific analysis setting.

Simulation-based analysis in DDDS can take different forms. The following are two commonly used simulation-based analysis:

- **What-if analysis** inspects how a system would behave if certain changes happened to the system. To carry out what-if analysis, one would modify a simulation model based on the changes that need to be tested and then run simulations to inspect the system's behavior under those changes.

- **Sensitivity analysis** determines how different values of an element (e.g., a model parameter or an external input) of a simulation model affect the simulation results. To carry out sensitivity analysis, one would systematically vary the values of the selected model element to run simulations and then assess the sensitivity of the simulation results to the selected model element.

5.8   Relation to Other Modeling and Simulation Activities

The new paradigm of DDDS is related to other modeling and simulation activities in various ways. Figure 5.10 shows how DDDS relates to modeling and offline simulation from a modeling and simulation lifecycle point of view. In the figure, the modeling activities include all the activities for developing a simulation model, such as conceptual modeling, simulation model implementation, and verification and validation. After a simulation model is developed, there are two ways of using the simulation model: offline simulation and DDDS. The offline simulation runs independently of a real system. This is the traditional way of simulation, mainly to support offline planning or system design. The DDDS runs concurrently with the real system. It is a type of online simulation but goes beyond the conventional online simulation by continuously and systematically assimilating real-time data. The goal of DDDS is to support real-time operation or decision-making.

The relationship between modeling and DDDS has two important implications for the quality of the simulation models used in DDDS. On the one hand, DDDS recognizes that simulation models are not perfect. It addresses the discrepancies between a simulation model and a real system by continuously assimilating real-time data. On the other hand, the new paradigm of DDDS does not diminish

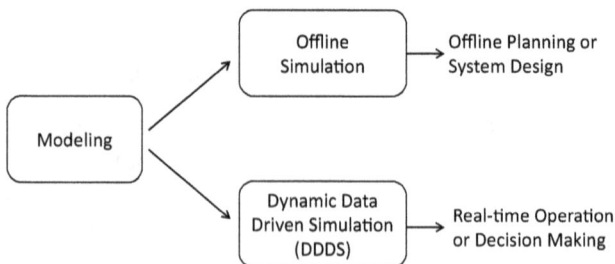

Figure 5.10.   DDDS and other modeling and simulation activities.

Table 5.1. Comparing offline simulation and DDDS.

	Offline simulation	DDDS
Application context	Supports offline planning or system design	Supports real-time decision-making or real-time operation
Experiment design	Simulates representative or hypothetical scenarios to study potential outcomes	Simulates the actual situation of a real system to provide real-time prediction/analysis
Simulation model configuration	Model is configured based on "typical" settings or hypothetical scenarios; multiple sets of parameter values may be used corresponding to the multiple experiment scenarios	Model is aligned with the real system in operation; parameter values may need to be estimated based on real-time measurement data collected from the real system
Data used	Historical data or hypothetical data	Real-time data from the real system
Main activities	Experiment design; simulation execution; offline analysis of simulation results	Dynamic state estimation; online model calibration; external input modeling & forecasting; simulation-based prediction/analysis
Execution time	Typically, there is no constraint on the execution time	Simulation executions need to be finished in real time

the importance of modeling because a high-quality simulation model is still essential. When the underlying simulation model is of low quality, it not only brings challenges to data assimilation but also directly impacts the results of simulation-based prediction/analysis.

After a simulation model is developed, offline simulation and DDDS run simulations in different ways. Table 5.1 compares the two ways of running simulations.

Despite major differences between offline simulation and DDDS, the two can also work together for the same application. Figure 5.11 shows how the two types of simulation can work together for the wildfire management application. Before a wildfire season, offline simulation is used to support the planning of firefighting resources. The simulation activity is to simulate potential wildfire ignition and spread scenarios in a study area to better understand the wildfire risk. During the wildfire season, DDDS is used to support real-time

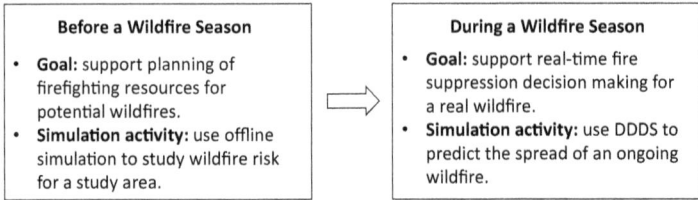

Figure 5.11. Offline simulation and DDDS work together for wildfire management.

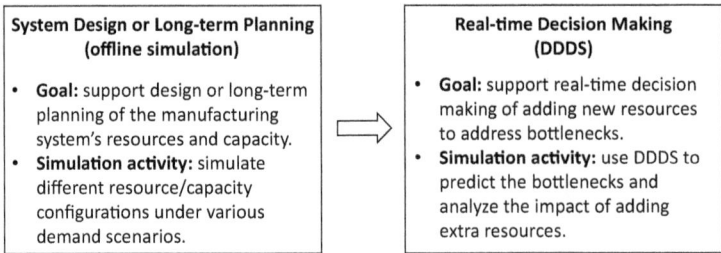

Figure 5.12. Offline simulation and DDDS work together for a manufacturing system.

decision-making for an existing fire. The simulation activity is to simulate and predict the spread of a real wildfire that is spreading.

Figure 5.12 shows another example where the two types of simulation work together for a manufacturing system application. In this application, the offline simulation is used to provide design or long-term planning for the manufacturing system's resources/capacity. The simulation activity is to simulate different resource and capacity configurations under various job demand scenarios. DDDS is used to support real-time decision-making when adding new resources to handle existing or predicted bottlenecks. The simulation activity is to simulate/predict the bottlenecks in the near future and analyze the impact of adding extra resources to address the bottlenecks.

5.9   Sources

The discussion of the limitations of data in Section 5.4 is influenced by Lahoz *et al.* (2010). The online model calibration in Section 5.5 is motivated by Zhang (2020). The description of time series analysis in Section 5.6.3 is influenced by Brownlee (2017).

Chapter 6

Data Assimilation

6.1 Introduction

The DDDS activities of *dynamic state estimation* and *online model calibration* use real-time measurement data to estimate a system's state and to calibrate a model's parameters. Both these activities can be framed as data assimilation problems and addressed by data assimilation methods.

Data assimilation is a methodology that combines measurement data with a dynamic model of a system to optimally estimate the evolving state of the system. Data assimilation was originally developed in the field of meteorology to provide weather forecasts using numerical weather models and measurement data of weather conditions. It has since gained popularity in many other science and engineering fields, such as geosciences, oceanography, hydrology, and robotics. This chapter describes data assimilation within the context of dynamic data-driven simulation, with a special interest in data assimilation for discrete simulations. Examples of discrete simulations include discrete time simulations, discrete event simulations, and agent-based simulations.

The work of data assimilation can be classified into *variational data assimilation* and *sequential data assimilation*. Variational data assimilation treats the estimation problem as an optimization problem — it finds solutions to minimize an *objective function* that measures the misfit between a sequence of model states and the measurement data that are available over a given assimilation window. Examples of variational data assimilation include three-dimensional variational assimilation (3DVAR) (Lorenc, 1986) and four-dimensional variational assimilation (4DVAR) (Courtier *et al.*, 1994). On the other hand, sequential data assimilation assimilates data sequentially with the objective of correcting the state estimate each time when a new measurement becomes available. Examples of sequential methods include Kalman-filter-based data assimilation and particle-filter-based data assimilation. The material in this book focuses on sequential data assimilation.

Data assimilation is a relatively new topic for the modeling and simulation field. To help grasp this new topic, this chapter starts with the fundamental concepts related to probabilistic state representation. Afterward, it proceeds to the formulation of the data assimilation problem and then describes the data assimilation methods. Throughout the descriptions, special attention is paid to the connections between data assimilation and simulation and to the treatments that are needed for working with discrete simulation models. Data assimilation deals with statistical estimation and is rooted in probability theory. The material in this chapter focuses more on the concepts and procedures of data assimilation as opposed to the theoretical underpinnings (e.g., mathematical derivations) of data assimilation methods.

6.2   State Space and Belief Distribution

The concept of *state* lies at the center of dynamic systems and is one of the key elements of a simulation model. To simulate a system's behavior in real time, a simulation needs to be initialized to the same (or a similar) state as that of the real system.

A fundamental premise of data assimilation is that our knowledge about a system's true state is imperfect. The precise value of the true state is unknown and uncertainty always exists. This holds true even when measurement data are available (see Chapter 5 for the limitations of measurement data). Building on this premise, data assimilation takes a probabilistic view of a system's state. Instead of working with states that have definitive values, data assimilation treats states as random variables that are expressed by probability distributions.

This book uses the term *belief* to refer to our (imperfect) knowledge about the true state. A belief is expressed through a probability distribution (also called a *belief distribution*) and is distinguished from the true state itself. A belief distribution assigns a probability (or density value) to each possible hypothesis with regard to the true state. The task of data assimilation is to estimate the belief distribution based on measurement data and a dynamic model of a system. Since the true state is unknown, the belief of the state then acts as the true state when needed, for example, to compute the expected value (i.e., the mean) of the true state or to initialize simulation runs in simulation-based prediction/analysis. This probabilistic view is different from how state is handled in simulations because simulation models generally work with states with definitive values.

A system's state is modeled by state variables. A state variable can be *continuous* or *discrete*. The latter can be further differentiated between *discrete scalar state variable* and *discrete categorical state variable*. The three types of state variables are described as follows:

- **Continuous state variable:** A numerical state variable that can assume an infinite number of real values within a given interval. Examples: the amount of water in a reservoir, an agent's horizontal or vertical position in a 2D continuous space.
- **Discrete scalar state variable:** A numerical state variable that can assume only discrete values. Examples: the number of jobs waiting in a queue, an agent's horizontal or vertical position in a discretized 2D cellular space.
- **Discrete categorical state variable:** A non-numerical state variable that describes an unquantifiable characteristic. Examples: the "busy" or "idle" state of a machine in a manufacturing system; the "unburned," "burning," or "burned" state of a forest cell in a wildfire spread simulation.

The full set of state variables forms a *state vector*, the dimension of which is determined by the number of state variables that are included. Depending on the types of state variables, a state vector defines different types of state. A state vector containing only continuous state variables defines a *continuous state*. A vector containing only discrete state variables defines a *discrete state*. A vector containing both continuous and discrete state variables defines a *hybrid state*. An example of a hybrid state is the state of a reservoir that operates in two modes: *open* and *closed*. In this example, the state vector includes a discrete categorical state variable describing the operating mode (i.e., *open* or *closed*) and a continuous state variable describing the amount of water inside the reservoir.

A *state space* is the set of all possible states of a state vector. The size of a state space depends on the possible values that each state variable can take on. In general, the state space of a continuous state or a hybrid state has an infinite size due to the infinite number of values that the continuous state variables can take on. For a discrete state, when all the state variables assume a finite number of values, its state space has a finite size, which can be calculated as the product of the numbers of possible values of all the state variables. Otherwise, when one or more state variables takes an infinite number of possible values (e.g., the number of jobs in an unbounded queue), the state space has an infinite size.

It is important to note that the belief distribution of a state vector is defined over its entire state space. In other words, the distribution assigns a probability (or density value) to each possible value in the state space.

6.3   Probabilistic State Representation

The fact that data assimilation works with beliefs of states means that it needs a way to represent belief distributions. In theory, a belief may be expressed in an arbitrary probability distribution that best describes the state of knowledge with regard to a state. An arbitrary probability distribution, however, often makes the data assimilation computation intractable. To achieve computationally feasible solutions, the belief distributions in data assimilation are often approximated by simplified representations. The two commonly used representations are:

- Gaussian representation;
- Sample-based representation.

The former works with continuous states and is the foundation for Kalman-filter-based data assimilation. The latter works with both continuous and discrete states and is the foundation for particle-filter-based data assimilation. These two representations also lead to different approaches to computing the belief distribution resulting from a state transition.

6.3.1 Gaussian representation

The Gaussian representation approximates a belief as a *Gaussian distribution* (also known as a *normal distribution*) that is fully described by a mean and a variance. A Gaussian distribution is a continuous probability distribution that works for continuous random variables. This means the Gaussian representation is applicable only to continuous states.

To demonstrate the principle of Gaussian representation, we consider the simplest case where the state is defined by a single continuous state variable. An example of this would be a vehicle's position on a one-dimensional road, where the only state variable is the vehicle's position, denoted by x ($x \in R$), where R is the set of real numbers. The state space of this example can be illustrated by a number line that includes all the real numbers. Assuming that at a specific time t, the vehicle's position (denoted by x_t^{true}) is at 78.5, i.e., $x_t^{\text{true}} = 78.5$. This is the vehicle's true state. Without knowing the true state, data assimilation treats the vehicle's position as a random variable (denoted by x_t) and expresses it using a probability distribution.

In the Gaussian representation, the belief of the vehicle's position is approximated by a Gaussian distribution, whose probability density function is defined in Chapter 4 (Equation (4.8)) and is restated in the following for convenience:

$$p\left(x_t\right) = \frac{1}{\sigma\sqrt{2\pi}}e^{-\frac{1}{2}\left(\frac{x_t-\mu}{\sigma}\right)^2}, \tag{6.1}$$

where μ is the mean of the distribution and σ^2 is the variance. The Gaussian distribution can be denoted by $x_t \sim N(\mu, \sigma^2)$.

Figure 6.1 shows an example of a Gaussian distribution to express the belief of the vehicle's position. In this example, the mean (μ) is 78.2, which is based on the fact that the vehicle's GPS sensor returns a value of 78.2. The variance (σ^2) is equal to 1.0, which is determined according to the precision measure of the GPS sensor. The probability density across the one-dimensional state space for

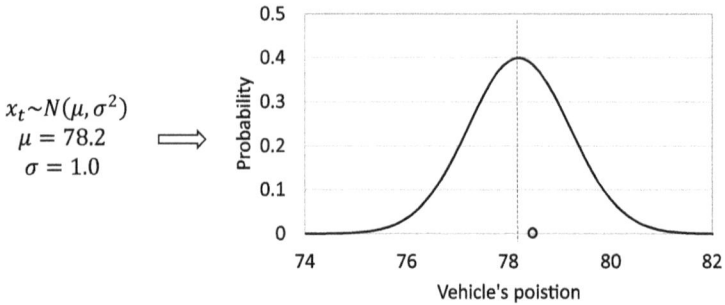

Figure 6.1.   Gaussian representation of a one-dimensional continuous state. The true state is marked as a circle on the horizontal axis.

this Gaussian distribution is shown on the right-hand side of the figure. The figure also marks the true position $x_t^{\text{true}} = 78.5$ as a circle on the horizontal axis. Since the true position is unknown, the Gaussian distribution $N(78.2, 1.0)$ represents the belief of the true state.

More generally, for an n-dimensional state vector, the Gaussian representation would use a *multivariate Gaussian distribution*, whose probability density function is defined in Chapter 4 (Equation (4.25)) and is restated as follows:

$$p(x_t) = \frac{exp\left(-\frac{1}{2}(x_t - \mu)^T \Sigma^{-1}(x_t - \mu)\right)}{\sqrt{(2\pi)^n |\Sigma|}}, \tag{6.2}$$

where x_t is an n-dimensional state vector at time t, μ is an n-dimensional mean vector, and $\sum$ is an $n \times n$ covariance matrix. The multivariate Gaussian distribution is denoted by $x_t \sim N(\mu, \sum)$.

The Gaussian representation has the advantage that it specifies belief distributions in an efficient manner, i.e., using only two parameters (μ and $\sum$). Furthermore, when working with models that are linear and have only Gaussian noises, the updates of the belief distributions remain Gaussian. This greatly simplifies the handling of the probability distributions and makes it possible to derive analytical solutions for the data assimilation problem (more details later). Another reason that makes the Gaussian representation a popular choice is that the probability distributions of many processes in science and engineering are indeed Gaussian or Gaussian-like, i.e., the probability densities concentrate on one specific value of the random variable and decrease according to a bell-shaped curve for other values. For example, the measurement errors of many sensors typically follow a zero-mean Gaussian distribution.

On the other hand, when the underlying distribution is not Gaussian (e.g., a multimodal distribution), using the Gaussian representation to approximate the distribution will lead to large errors. The Gaussian representation also has several other limitations. In particular, it does not work for discrete states, which are used by many discrete simulation models. It also depends on linear Gaussian models (described later) in order to maintain the Gaussian property when updating belief distributions.

6.3.2 *Sample-based representation*

The sample-based representation approximates the belief of a state using a set of samples. Each sample is a concrete instantiation of the state, i.e., a hypothesis as to what the true state may be. The law of large numbers states that if the sample size is large enough, a desired level of accuracy can be achieved for representing any arbitrary probability distributions. Let $\{x_t^{(i)}; i = 1, \ldots, N\}$ be the sample set at time t holding samples of a target distribution, then the probability density function of the target distribution can be approximated as

$$p(x_t) \approx \frac{1}{N} \sum_{i=1}^{N} \delta(x_t - x_t^{(i)}), \qquad (6.3)$$

where $p(x_t)$ is the probability density function (computed based on $\{x_t^{(i)}; i = 1, \ldots, N\}$) for the random variable x_t, N is the size of the sample set, and $\delta()$ is the Dirac delta function.

Figure 6.2(a) illustrates a sample-based representation for the belief of a mobile agent's position in a 2D continuous space. In this example, the state vector includes two continuous state variables, denoted by $x_t = [x_{1,t}, x_{2,t}]^T$, where $x_{1,t}$ and $x_{2,t}$ stand for the horizontal position and vertical position of the agent, respectively, at time t, and ^{T} is the transpose notation transposing a row vector to a column vector. Figure 6.2(a) shows 200 samples, each of which represents a hypothesis of the agent's position in the 2D space. For comparison purposes, Figure 6.2(b) shows a corresponding Gaussian representation for the belief of the agent's position. The two-variate

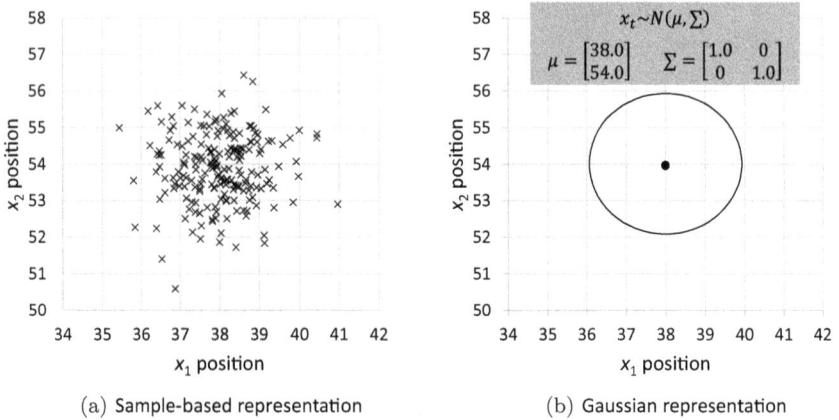

(a) Sample-based representation　　　　(b) Gaussian representation

Figure 6.2.　(a) Sample-based representation and (b) a corresponding Gaussian representation for a 2D continuous state vector.

Gaussian distribution is visualized using its mean and the covariance error ellipse that contains 95% of the samples.

Besides working with continuous states, the sample-based representation also works with discrete and hybrid states. When working with discrete states, the samples are drawn from the state space of the discrete state variables. When working with hybrid states, each sample is a state vector drawn from the state space including both continuous and discrete state variables.

Figure 6.3 illustrates a sample-based representation for the belief of a discrete scalar state that models the number of jobs (referred to as *queue length*) in a queue with a capacity of 10. The figure shows the frequency diagram of 200 samples. A frequency diagram can be converted to a probability distribution diagram by calculating the ratio of the frequency values over the total sample size (200). In this example, it can be seen that there are roughly equal probabilities for the queue to be close to empty (centered around queue length = 1) and close to full (centered around queue length = 9), and there is zero probability for the queue length to be 4, 5, or 6. A probability distribution like this is an example of a multimodal distribution that has more than one peak, or "modes."

Figure 6.4 illustrates a sample-based representation for the belief of a hybrid state used in the tutorial example to be described in Chapter 7. The hybrid state vector in this example includes five state

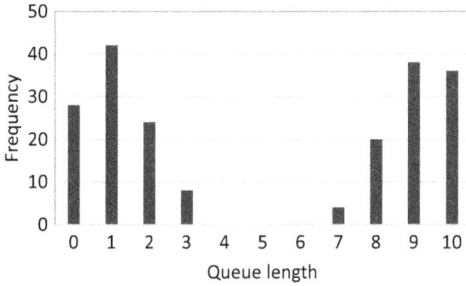

Figure 6.3. Sample-based representation for a discrete scalar state. In this example, the state space is defined by the integer values between 0 and 10 (inclusive). The sample set includes 200 samples. The frequency diagram of the samples' values is displayed (for simplicity, the actual values of the 200 samples are omitted).

variables $[x_{trafficLightState}, x_{sigma}, x_{elapsedTimeInGreen}, x_{westQueue},$ $x_{eastQueue}]^T$. $x_{trafficLightState}$ is a discrete categorical state variable describing the states of the traffic light, which can be either *westMovGreen* or *eastMovGreen*, indicating which direction has the green signal. x_{sigma} and $x_{elapsedTimeInGreen}$ are continuous state variables describing the remaining time and elapsed time, respectively, in the current traffic light state, and they are non-negative real numbers. $x_{westQueue}$ and $x_{eastQueue}$ are discrete scalar states describing the number of vehicles waiting in the west- and east-side queues, respectively, and they have non-negative integer values. In a sample-based representation, each sample would be a state vector having specific values for all the five state variables. For simplicity, Figure 6.4 shows only two of the state variables: $x_{trafficLightState}$ and $x_{elapsedTimeInGreen}$.

Figure 6.4 shows the histogram of 1,000 samples, which represents the probability distribution of the state belief. The histogram indicates that the traffic light is most likely in the *eastMovGreen* state and has been in that state for a relatively long time (because $x_{elapsedTimeInGreen}$ of the corresponding samples are all in the range of [70, 93] s). There is also a small possibility for the traffic light to be in the *westMovGreen* state for a short period of time (in the range of [1, 15] s). The probability for the traffic light to be in the *eastMovGreen* state is much larger than that of the *westMovGreen* state due to the much larger number of samples (817 vs. 183) representing the former.

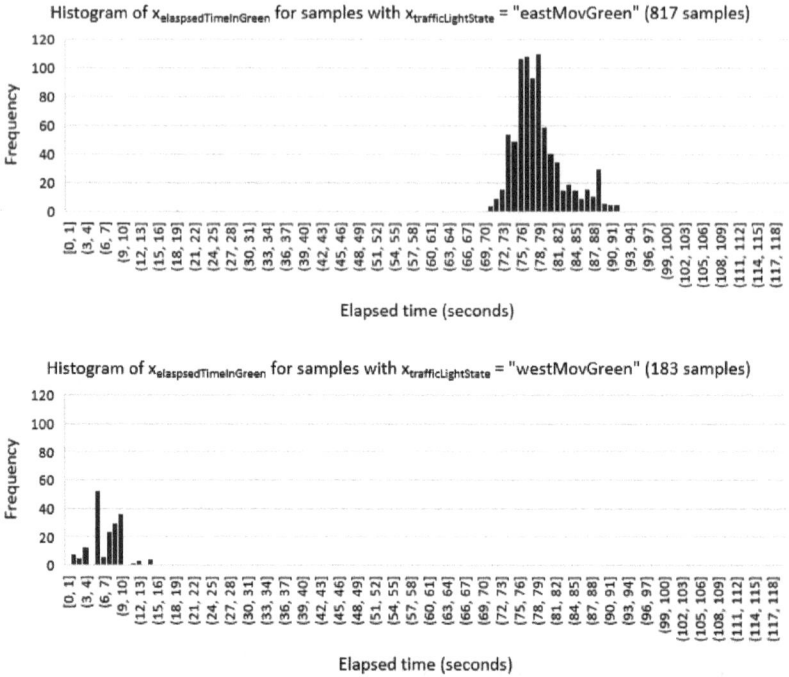

Figure 6.4. Sample-based representation for a hybrid state. The sample set has 1,000 samples, among which 817 samples' $x_{trafficLightState}$ has the value of $eastMovGreen$, and 183 samples' $x_{trafficLightState}$ has the value of $westMovGreen$. The top figure shows the histogram of $x_{elapsedTimeInGreen}$ for the 817 samples whose $x_{trafficLightState} = eastMovGreen$; the bottom figure shows the histogram of $x_{elapsedTimeInGreen}$ for the 183 samples whose $x_{trafficLightState} = westMovGreen$.

The sample-based representation has the advantage that it can approximate any arbitrary distribution as long as the sample size is sufficiently large. This is different from the Gaussian representation that results in large errors when approximating non-Gaussian distributions. Furthermore, a major advantage of the sample-based representation is that it works with discrete and hybrid states, and it supports nonparametric approaches to updating belief distributions. In a sample-based representation, each sample is an individual participating in computations, and the overall probability distribution is defined by the results of all the samples. This makes it possible to work with discrete simulation models that have nonlinear non-Gaussian behavior. On the negative side, the sample-based representation is not as efficient as the Gaussian representation because it involves using a large number of samples.

6.3.3 *Updating belief distribution from state transition*

A major characteristic of dynamic systems is that their states change over time. In data assimilation, the knowledge about how a system evolves its state is captured by a dynamic model (called the *state transition model*) that defines the state transition of the system. Because data assimilation works with belief distributions of states, the beliefs after state transitions need to be expressed in probability distributions too. In other words, given an initial belief of a state before a state transition, there is a need to compute the belief of the resulting state after the state transition. The resulting belief depends on the initial belief and the dynamic model that governs the state transition.

Depending on the representation of the initial state belief and the properties of the dynamic model, one can use different approaches to compute and approximate the resulting belief after a state transition. In the following, we first describe a *sampling* approach that works for general dynamic models and then describe an analytical approach that works for a special type of dynamic models called *linear Gaussian models*. The sampling approach uses the sample-based representation, and the analytical approach uses the Gaussian representation.

It is important to note that computing the probability distribution from a state transition is only one of the steps of data assimilation. Another key step is to update the computed probability distribution using information from the measurement data. The procedure of data assimilation will be described later.

6.3.3.1 *The sampling approach*

The sampling approach is to draw a set of samples using the dynamic model to approximate the belief distribution of a state after a state transition. This approach works naturally with the sample-based representation. In the sample-based representation, the belief distribution is represented by a set of samples. Each sample is a concrete instantiation of the state and can be treated independently to participate in a computation, e.g., a state transition. Given a dynamic model and an initial set of samples, the sampling approach generates a new set of samples in the following way: For each sample in the initial sample set, it advances the state represented by the sample to a new state value using the state transition model. After all the samples finish the state transition, the full set of new states becomes the new set of samples to represent the belief distribution of the state after the state transition. The sampling approach is a case of the Monte Carole method described in Chapter 4.

A simulation model is a dynamic model that models how a system's state changes over time. The sampling approach thus works well with all simulation models. When working with a simulation model, for each sample, the sampling approach uses the state defined by that sample as the initial state and runs a simulation for the period of time that is defined by the forward time window of the state transition. The resulting states from all the samples then form the new sample set.

Figure 6.5 illustrates the sampling approach using the mobile agent example shown in Figure 6.2. In this example, the mobile agent moves in a 2D continuous space. Its state is defined as $x = [x_{1,t}, x_{2,t}]^T$, where $x_{1,t}$ is the horizontal position and $x_{2,t}$ is the vertical position at time t. The mobile agent's movement is modeled by the dynamic model:

$$dx_{1,t}/dt = 6.0 + \gamma_{x1},$$
$$dx_{2,t}/dt = 8.0 + \gamma_{x2}, \qquad (6.4)$$

where 6.0 (m/s) and 8.0 (m/s) are the agent's moving speeds along the horizontal and vertical directions, respectively, and $\gamma_{x1} \sim N(0, 4.0)$ and $\gamma_{x2} \sim N(0, 4.0)$ model the process noises (e.g., due to motor noises or uneven road conditions) of the movement. γ_{x1} and γ_{x2} are zero-mean Gaussian noises and are independent of each other. This dynamic model is a continuous simulation model with stochastic noises added.

Let us assume that the belief of the agent's position at a time t is represented by a sample set of 200 samples, displayed in the left diagram of Figure 6.5. We are interested in computing the belief distribution of the agent's new position after it moves for 1 s ($\Delta t = 1\,s$).

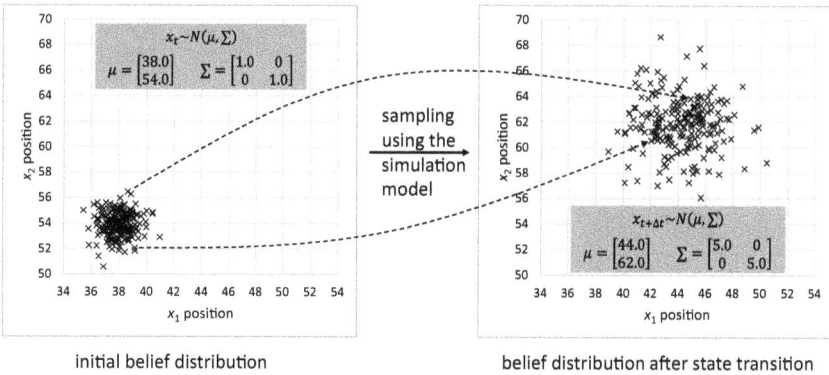

Figure 6.5. The sampling approach for computing probability distribution after a state transition. The initial belief is represented by 200 samples displayed in the left diagram. These samples are simulated to generate 200 new samples displayed in the right diagram. The 200 new samples represent the belief of the agent's new position after the state transition. The relationship between two pairs of old and new samples is illustrated by two arrows in the figure.

In other words, given the initial belief of the agent's state represented by the 200 samples, we want to compute the belief of the agent's state after the state transition that has a forward time window of 1 s.

For each sample in the sample set, the sampling approach simulates the changes in x_1 and x_2 for 1 s using the continuous simulation model defined in Equation (6.4), starting from the initial state defined by the sample. The resulting 200 samples are then used to represent the belief of the new state after the state transition. In this example, the initial samples are centered around the position $[38.0, 54.0]^T$. The new samples are centered around the position $[44.0, 62.0]^T$ and are more scattered compared to the initial samples. This is due to the process noises that are involved in the state transition. In fact, in this example, the initial 200 samples are drawn from a bivariate normal distribution with the mean and covariance matrix shown inside the gray box in the left diagram. Because the dynamic model is a linear Gaussian model (described later), the probability distribution of the resulting state still follows a bivariate normal distribution, whose mean and covariance matrix are shown inside the gray box in the right diagram.

The sampling approach is a general approach that works with the sample-based representation to compute the belief distribution after a state transition. This approach can be universally applied regardless of the state type (continuous, discrete, or hybrid), the dynamic model type (e.g., continuous simulation model or discrete simulation model), and the properties of the dynamic model (e.g., linear or nonlinear, Gaussian or non-Gaussian). As long as the belief distribution is represented by samples and a dynamic model is available, this approach works. Furthermore, this approach can be recursively applied in multiple iterations of a state transition because each iteration results in a new sample set that can be used in the next iteration.

6.3.3.2   *The analytical approach*

Having described the sampling approach that works for general dynamic models, we turn to a special, yet important, type of dynamic model where an analytical approach can be applied to update the belief distribution from a state transition. This analytical approach works with the Gaussian representation that approximates the beliefs of continuous states using Gaussian distributions.

The special type of dynamic models are models that include *linear* functions of state transition with added Gaussian noise. A linear function of a state transition means that the resulting state of the state transition is a linear combination of the state variables from the beginning state. Gaussian noise means that the uncertainty added to the state transition follows a Gaussian distribution. We name this type of dynamic model the *linear Gaussian model* or the *linear Gaussian state transition model*. A formal definition and treatment of the linear Gaussian model will be provided in Section 6.7.1.

A linear Gaussian model has an important consequence for updating the belief distribution resulting from a state transition. Specifically, when a state transition starts from a belief that has a Gaussian distribution, the belief of the resulting state will still follow a Gaussian distribution even though the mean and covariance may be changed. In other words, a linear Gaussian dynamic model preserves the Gaussian property of a belief. The preservation of the Gaussian property gives rise to an analytical approach for computing the belief distribution resulting from a state transition. In this analytical approach, the mean and covariance of the Gaussian distribution of the resulting belief can be analytically derived using arguments of the model as well as the mean and covariance of the initial belief

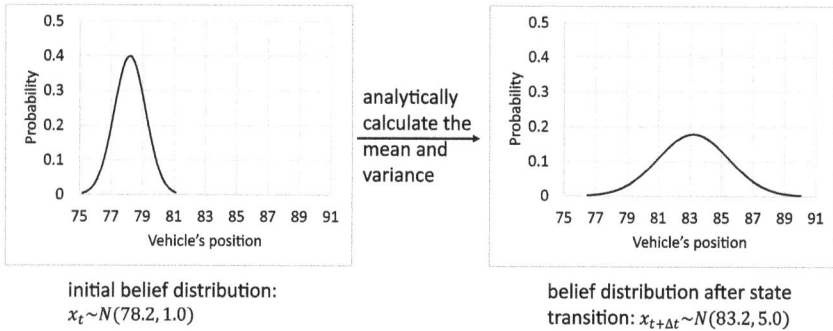

Figure 6.6. The analytical approach to computing belief distribution from a state transition.

of the state (for details, see Equations (6.39) and (6.40) in Section 6.7.1). This compares to the sampling approach that needs to run simulations for all the samples using the simulation model.

Figure 6.6 illustrates the analytical approach using the vehicle example shown in Figure 6.1. In this example, the vehicle moves on a one-dimensional road at a constant speed of 5.0 m/s. The vehicle's state (x_t) is its position, and the state transition model is defined by $dx_t/dt = 5.0 + \gamma$, where $\gamma \sim N(0, 4.0)$ is a Gaussian noise modeling the uncertainty of the vehicle's movement. This state transition model is a linear Gaussian model. Let us assume that at time t, the belief of the vehicle's position is described by the Gaussian distribution $x_t \sim N(78.2, 1.0)$. After the vehicle moves for 1 s (i.e., $\Delta t = 1$ s), the belief of the vehicle's new position would still be a Gaussian distribution. The mean and variance of the new Gaussian distribution can be analytically calculated: $\mu = 78.2 + 5.0 = 83.2$ and $\sigma^2 = 1.0 + 4.0 = 5.0$. Thus, the belief distribution of the vehicle's new state $(x_{t+\Delta t})$ after the state transition is $x_{t+\Delta t} \sim N(83.2, 5.0)$.

The analytical approach of updating the belief distribution resulting from a state transition works only when the state transition model is a linear Gaussian model. For nonlinear non-Gaussian models, the resulting belief distribution is no longer Gaussian. One may use approximation methods to work with these models. For example, in the extended Kalman Filter (EKF), a nonlinear state transition model would be linearized so that the linear approximation of the model can result in a Gaussian distribution. For more general cases, the sampling approach should be used, especially when a state transition model has strong nonlinear or non-Gaussian behavior. Also, note that the analytical approach works only for continuous states due to the Gaussian representation that it relies on.

6.4   The Data Assimilation Problem

Data assimilation deals with state estimation by combing information from the measurement data and a dynamic model. It generally adopts a *state-space formulation* and uses a probabilistic approach to formulate the state estimation problem.

Before looking into the formulation, we introduce the notation for defining the time base for a dynamic system. Let $t_0, t_1, \ldots, t_k$ (with $t_0 < t_1 < \cdots < t_k$) be a sequence of time instants, where the index $k \in N^0$ (N^0 is the set of non-negative integer numbers) is called the *time step*. The interval between two consecutive time steps, i.e., $\Delta t_k = t_k - t_{k-1}(i = 1, 2, \ldots)$, may not be constant. In the following description, we use the subscript k to refer to a variable's value at time t_k and subscript $i : k$ to refer to the sequence of values from time t_i to t_k. For example, we define $x_k := x(t_k)$ as the value of x at time t_k and $x_{0:k} := \{x_0, x_1, \ldots, x_k\}$ as the sequence of x's values from t_0 to t_k.

6.4.1   *State-space formulation*

In data assimilation, a dynamic system is generally formulated as a dynamic *state-space model*, which is composed of the *state transition model* of Equation (6.5) and the *measurement model* of Equation (6.6) as follows:

$$x_k = f_k\left(x_{k-1}, u_k, \gamma_k\right), \tag{6.5}$$

$$y_k = g_k\left(x_k, \varepsilon_k\right), \tag{6.6}$$

where x_{k-1} and x_k are the state vectors of steps $k - 1$ and k, respectively, u_k is the external input vector of step k, and y_k is the measurement vector of step k. The x_k, y_k, and u_k are all vertical vectors in the form of

$$x_k = \begin{bmatrix} x_{1,k} \\ x_{2,k} \\ \vdots \\ x_{n,k} \end{bmatrix}, \quad y_k = \begin{bmatrix} y_{1,k} \\ y_{2,k} \\ \vdots \\ y_{m,k} \end{bmatrix}, \quad \text{and} \quad u_k = \begin{bmatrix} u_{1,k} \\ u_{2,k} \\ \vdots \\ u_{l,k} \end{bmatrix}, \qquad (6.7)$$

where n, m, and l are the dimensions of the x_k, y_k, and u_k vectors and $x_{i,k}, y_{i,k}$, and $u_{i,k}$ are the ith elements of the x_k, y_k, and u_k vectors, respectively. The x_k, y_k, and u_k vectors may have different dimensions.

The function $f_k()$ defines the dynamics of the state transition, and the function $g_k()$ defines the mapping from the state to the measurement. γ_k and ε_k are two independent random variables modeling the noises (or uncertainties) involved in the state transition and measurement, respectively. The former is called the *process noise* and the latter is called the *measurement noise*. At step k, the sequences of external inputs $u_{1:k}$ (i.e., input data) and measurements $y_{1:k}$ (i.e., measurement data) up to this step are assumed to be known. The initial state x_0 is assumed to be known too based on some prior knowledge. Nevertheless, the states $x_{1:k}$ are hidden and cannot be observed directly. They need to be estimated.

The state transition model (Equation (6.5)) captures the knowledge about how the dynamic system evolves its state over time. This knowledge is important for data assimilation as it can generate a prediction of the belief of the state at a future time. A simulation model models a system's state transition and thus can serve as a state transition model. The measurement model (Equation (6.6)) links the state with the measurement data so that information from the latter can be utilized to update the (predicted) belief of the state. A measurement model can be viewed as the model of the sensors that collect the measurement data. It describes, at some level of abstraction, how the state x_k causes the sensor measurement y_k. As will be detailed later, the two models each play unique roles in estimating the dynamically changing state of a system.

Due to the probabilistic nature of state estimation, the state transition model (Equation (6.5)) is also called the *state transition density* and is expressed by $p(x_k|x_{k-1}, u_k)$; the measurement model (Equation (6.6)) is also called the *measurement density* and is

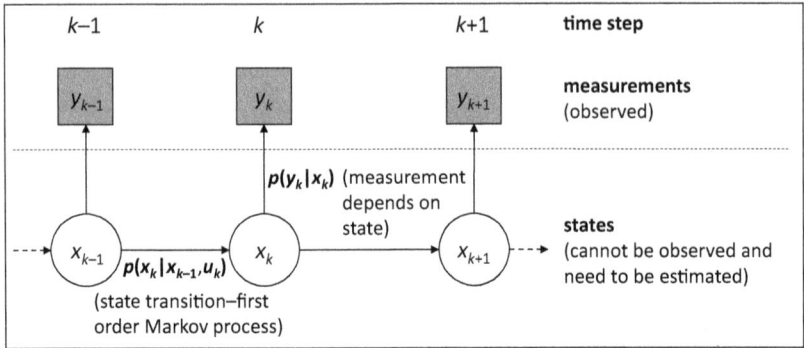

Figure 6.7.   States and measurements of a dynamic system.

expressed by $p(y_k|x_k)$. Thus, the state-space model can also be specified in the probability density form:

$$p(x_k|x_{k-1}, u_k), \tag{6.8}$$

$$p(y_k|x_k). \tag{6.9}$$

The probability density form of the state-space model makes it explicit that the state transition and the state–measurement mapping are probabilistic, resulting in probability distributions of the state and measurement at each step. This probabilistic view is necessary because a state is known only by its belief, and that uncertainties exist for both the state transition and measurement processes. Figure 6.7 shows the relationships among the state, measurement, state transition model, and measurement model.

6.4.1.1   *The Markov assumption*

The state-space formulation assumes that the state transition $p(x_k|x_{k-1}, u_k)$ is a first-order Markov process, meaning that the future state depends only on the present state as well as the future inputs that may be received. Based on the Markov assumption, the state x_{k-1} is a *complete* representation of the system at step $k-1$ for predicting the future state x_k. The knowledge of past states, external inputs, or measurements carries no additional information for predicting the future as they have been fully incorporated in x_{k-1}. In mathematical terms, this is expressed by the following conditional independence:

$$p(x_k|x_{0:k-1}, y_{1:k-1}, u_{1:k}) = p(x_k|x_{k-1}, u_k). \qquad (6.10)$$

The completeness of the state also means that the measurement y_k at step k is fully defined by the present state x_k; it does not depend on past states or past measurements. Mathematically, we have

$$p(y_k|x_{0:k}, y_{1:k-1}, u_{1:k}) = p(y_k|x_k). \qquad (6.11)$$

The Markov assumption plays a fundamental role in sequential data assimilation. With this assumption, the past and future can be treated independently as long as we know the present state. This makes it easier to implement data assimilation because there is no need to store and revisit past data once they are processed. All we need to keep track of is the belief of the most recent state.

The Markov assumption has practical implications for a simulation model used in data assimilation. To satisfy the complete state requirement, a simulation model should cover all the important dynamics of a system so that its state is a "complete" representation of the system, from which the future state can be predicted. Otherwise, the future state is not fully defined due to the incompleteness of the present state. This does not mean that every variable that may influence a system needs to be included in the simulation model. Instead, a model should be developed at the right abstraction level so that the unmodeled variables have close-to-random effects, which can then be treated as process noises. This principle of using the right state representation is consistent with the principle of developing simulation models at the right abstraction level.

6.4.1.2   *Simulation model for the state-space formulation*

The state-space formulation in Equations (6.5) and (6.6) defines the state transition model and measurement model, respectively, in a general way. It poses no restriction on the specific format of the models. In practice, the models for different applications may vary significantly. For example, for the linear Gaussian system to be described in Section 6.7.1, $f_k()$ and $g_k()$ are defined by mathematical equations. For the simulation applications considered in this book, $f_k()$ and $g_k()$ may be complex simulation models and measurement models defined by computer programs. This state-space formulation applies to all the models regardless of their formats.

Notice that the state transition model (Equations (6.5)) is defined in a way so that the state is updated in discrete time steps. This works naturally for continuous simulation models and discrete time simulation models because both can be implemented using a step-wise update mechanism. For discrete event simulation models, state updates are driven by the occurrences of external/internal events. The state transitions of a discrete event simulation can be formalized as an *integer indexed state process*, whose state representation includes the elapsed time since the last event. This allows a discrete event simulation model to be formulated in the same way as Equation (6.5) of the state-space formulation (Xie and Verbraeck, 2019). In general, a simulation model can serve as the state transition model in the state-space formulation.

6.4.1.3 *Process noise and measurement noise*

The process noise (γ_k) and measurement noise (ε_k) are essential components of the state transition model and measurement model. The necessity of incorporating noises in these models can be explained from two viewpoints. The first viewpoint focuses on the stochastic nature of the state transition and measurement data of the real system, which results in uncertainties that are best modeled using noises. For example, the spread of wildfire is a highly stochastic process due to the dynamic combustion reactions influenced by many factors. The process noise makes it possible to capture the uncertainties involved in a fire spread. Similarly, measurement data are inherently noisy due to factors such as background variability and sensor precision. For example, the sea level is inherently unstable, and thus, measurement data are noisy regardless of what type of sensors are used. The second viewpoint focuses on the fact that models are always imperfect. The noise factors provide a way to model the effects of the unmodeled variables. They help cover the discrepancies that may exist between the model output and the true value.

Adding process noise to a state transition may not be the first thought that comes to mind when developing a simulation model. This is especially true for systems that are often modeled in a deterministic way (e.g., a traffic light that switches regularly according to a fixed schedule). Nevertheless, in data assimilation, introducing noise to a deterministic model in the state-space formulation is still beneficial. The introduced noise is not to model the stochastic behavior of a system. Instead, it is to model the discrepancies that may exist between the model and the real system. The noise factor increases the chance for the belief of the predicted state to cover the true state so that more robust state estimation can be achieved when updating the predicted state using measurement data.

The scales of process noise and measurement noise should be informed by the uncertainties that need to be modeled. In general, the process noise should have a relatively small impact on the state update compared to that from the main dynamics modeled by the simulation model. Otherwise, the process noise would overshadow the simulation model when updating the state. For the measurement noise, the noise distribution should match the characteristics of the sensor measurement as well as the background variability. For example, when modeling a sensor's noise using a Gaussian distribution, the standard deviation σ needs to be defined based on the precision of the sensor: A sensor with high (low) precision would have a small (large) σ. Note that the process noise and measurement noise may not be Gaussian for many applications.

It is important to understand the roles of the process noise and measurement noise in state estimation. The two noises carry information about the confidence of the model prediction and sensor measurement, respectively. The relative magnitude between them determines how much trust to put in the information obtained from the two. Intuitively, if the process noise is large and the measurement noise is small, it makes sense for the state estimation to rely more on the information from the measurement data. If the opposite is true, the state estimation should rely more on the model prediction. The Bayesian approaches described later provide systematic ways to incorporate the information from the two sources. For example, in the Kalman filter, a Kalman gain is calculated based on the covariances of the process and measurement noises and then used as a weight factor for combining the information from the model prediction and measurement data. One of the key advantages of data assimilation is that it works with imperfect models and noisy measurements in a systematic way.

6.4.2 *Filtering and smoothing*

Based on the state-space formulation, the task of state estimation is to produce an accurate estimate of the state of a dynamic system based on measurement data. This can be formulated as two different problems by estimating two posterior distributions related to the state:

- $p(x_k|y_{1:k}, u_{1:k})$: estimating the distribution of the state x_k at the present time given all of the measurement data and external inputs up to the present time. This is referred to as the *filtering* problem.
- $p(x_n|y_{1:k}, u_{1:k})$, where $n = 1, \ldots, k$: estimating the distribution of the state at a previous time given all of the measurement data and external inputs up to the present time. This is referred to as the *smoothing* problem.

The filtering problem estimates the present state of a system based on the measurement data received up to the present time. It works naturally in a real-time setting, i.e., estimating the new real-time state whenever new measurement data arrive. The smoothing problem estimates the state at a previous time given all of the measurement data up to some later time. It is called smoothing because the estimated state trajectories, as a result of the additional information available, tend to be smoother than those obtained by filtering. The smoothing can achieve better estimation results because it utilizes later observations that are not available in the filtering problem.

A similar formulation can be used to define the *prediction* problem, i.e., estimating $p(x_n|y_{1:k}, u_{1:k})$ for $n > k$. The prediction problem estimates the state at a future time given all of the measurement data and external inputs up to the present time. The prediction problem may be addressed by first solving the filtering problem, i.e., estimating the present state x_k, and then starting from the present state x_k and using the state transition model $p(x_{k+1}|x_k, u_{k+1})$ recursively to predict the future states, assuming that the future inputs $u_{k+1:n}$ are known.

The material in this book focuses only on the filtering problem. Strategies for addressing the smoothing problem can be found in the literature, such as Briers *et al.* (2010) and Doucet and Johansen, (2009).

6.4.3 *Joint state–parameter estimation*

Our discussion so far assumes that the state transition model $p(x_k|x_{k-1}, u_k)$ is given with all its model parameters known. The model parameters are not explicitly listed in the state transition model because they are a fixed part of the model. This is true in many cases, as model parameters are calibrated offline to have definitive values before a model is used. Nevertheless, situations also exist when one needs to calibrate some model parameters in an online fashion (see discussions in Section 5.2). In these situations, there is a need to estimate the corresponding model parameters dynamically based on real-time measurement data.

A common approach for online parameter estimation is to formulate it as a joint state–parameter estimation problem. In this approach, the to-be-estimated parameters are included as part of the state vector that needs to be estimated. Let x_k be the n-dimensional state vector and θ_k be the h-dimensional parameter vector that need to be estimated at step k. Typically, the set of parameters that need to be estimated is a small subset of all the parameters of a dynamic model. We define an augmented state vector z_k by appending the parameter vector θ_k to the state vector x_k, i.e.,

$$z_k = \begin{pmatrix} x_k \\ \theta_k \end{pmatrix} \text{ or } z_k = (x_{1,k}, x_{2,k}, \ldots, x_{n,k}, \theta_{1,k}, \theta_{2,k}, \ldots, \theta_{h,k})^T, \quad (6.12)$$

where z_k is an $n + h$ dimensional vector, $x_{i,k}$ $(i = 1, \ldots, n)$ is the ith element of the state vector, and $\theta_{j,k}$ $(j = 1, \ldots, h)$ is the jth element of the parameter vector.

The dynamics of the state x_k is defined by the state transition model (Equation (6.5)). To formulate the state-space model for the joint state–parameter estimation, we need a way to define how the parameters θ_k evolve over time. A popular treatment is to add small random perturbations to the parameters in each step of the transition (Liu and West, 2001). Typically, the random perturbations are modeled by zero-mean Gaussian distributions with some specified variances for each parameter element, i.e.,

$$\theta_k = \theta_{k-1} + \zeta_k, \tag{6.13}$$

$$\zeta_k \sim N(0, W_k), \tag{6.14}$$

where ζ_k are the random perturbations modeled as zero-mean Gaussian noises and W_k is a diagonal covariance matrix for the Gaussian noises that has a variance of σ_j^2 for the jth parameter. The initial parameter values can be set based on the information from offline model calibration. Adding random perturbations to the parameter values allows for the generation of new parameter values at each step of data assimilation. This supports robust estimation even if the initial values are not close to the true values. Large variances in the Gaussian noises lead to large changes in the parameter values in each step. The large variances may be needed if the estimation has not converged or if one expects the parameter values to change dynamically at a fast pace. Otherwise, small variances are preferred.

Combining the parameters' dynamic model Equation (6.13) with the state transition model Equation (6.5), we have

$$z_k = \tilde{f}_k \left(z_{k-1}, u_k, \gamma_k, \zeta_k \right) = \begin{pmatrix} f_k(x_{k-1}, \theta_{k-1}, u_k, \gamma_k) \\ \theta_{k-1} + \zeta_k \end{pmatrix}. \qquad (6.15)$$

Note that the $f_k()$ function in Equation (6.15) explicitly lists the model parameters θ_{k-1} because the state transition of step k should use the estimated model parameters from step $k-1$. In joint state–parameter estimation, the model parameters at different steps need to be differentiated because their values vary over time.

With the augmented state vector z_k, the measurement model can be rewritten as

$$y_k = \tilde{g}_k \left(z_k, \varepsilon_k \right) \stackrel{\text{def}}{=} g_k \left(x_k, \varepsilon_k \right), \qquad (6.16)$$

where $\tilde{g}_k()$ maps from the augmented state vector z_k to the measurement vector y_k. Here, $\tilde{g}_k \left(z_k, \varepsilon_k \right)$ is equivalent to $g_k \left(x_k, \varepsilon_k \right)$ because y_k is completely defined by the state x_k.

Equations (6.15) and (6.16) together form the state-space model for the joint state–parameter estimation problem. Based on this formulation, the parameters and the state can be jointly estimated using the same data assimilation methods to be described later. An example of joint state–parameter estimation is provided in Chapter 7.

6.5   Sequential Bayesian Filtering

Having defined the data assimilation problem, it remains to show how to estimate the dynamically changing state using measurement data. This book focuses on *sequential data assimilation* that provides state estimation sequentially when new measurement data become available. A popular framework for sequential data assimilation is *sequential Bayesian filtering*. Both the Kalman filter and particle filters, which will be described later, are based on the sequential Bayesian filtering framework.

The key feature of sequential Bayesian filtering is that each step of the filtering utilizes the state estimate from the previous step to predict and update a new state estimate, resulting in a recursive procedure that can be applied whenever new measurement data become available. This recursive procedure gives rise to a stepwise implementation that can be carried out in real time. Each step involves a prediction of a new state belief using the state transition model and an update of the predicted belief using the measurement data. The updated belief represents the new state estimate and serves as the starting point for the next-step state estimation.

6.5.1   *Mathematical derivation*

We first provide a mathematical derivation of the sequential Bayesian filtering, focusing on how a recursive procedure can be established. The derivation also shows how the state transition model and measurement model play their roles in computing the state estimate in each step. The derivation in this section is adapted from Thrun *et al.* (2005). For simplicity, we assume all the random variables and distributions are continuous. A similar derivation can be done for discrete random variables. Such a derivation would lead to the same final results and is omitted here.

To get started, we first revisit the Bayes theorem, which describes how new evidence (also called *data*) can be incorporated to update the belief of a random variable. Mathematically, the Bayes theorem is stated by the following equation:

$$p\left(x|y\right) = \frac{p\left(y|x\right)p(x)}{p(y)}, \tag{6.17}$$

where x is a random variable that we would like to infer based on the data y. This is expressed by the conditional probability $p\left(x|y\right)$ on the left-hand side of the equation, which is generally called the *posterior distribution* (or simply *posterior*) of x. On the right-hand side of the equation, the probability $p(x)$ is called the *prior distribution* (or simply *prior*), which summarizes our knowledge about x prior to incorporating the data y. The conditional probability $p(y|x)$ is called the *likelihood*, which describes the probability of observing the data y if x is true. Finally, the probability $p(y)$ is often called the *marginal likelihood*. The marginal likelihood $p(y)$ is independent of x. This means the factor $p(y)^{-1}$ will be the same for any value x in the posterior $p(x|y)$. The role of $p(y)^{-1}$ in Equation (6.17) is to normalize the probability values so that the posterior $p(x|y)$ integrates to 1. For this reason, $p(y)^{-1}$ is often written as a *normalizer* factor and is denoted by η. Then, the Bayes Equation (6.17) can be rewritten as

$$p\left(x \mid y\right) = \eta p\left(y \mid x\right)p(x). \tag{6.18}$$

Equation (6.18) shows that the posterior distribution $p\left(x \mid y\right)$ is proportional to the prior distribution $p(x)$ multiplying the likelihood $p(y|x)$. Note that the likelihood $p(y|x)$ is the "inverse" conditional probability of the posterior $p(x|y)$. This means that if we are interested in estimating a random variable x after observing data y, the Bayes theorem provides a way to do so by using the inverse probability that specifies the probability of observing the data y assuming that x was true. This is convenient because the measurement model in the state-space formulation defines the mapping from the state x to the measurement data y and thus provides information for computing the likelihood.

In the following, we show how a recursive formula can be established so that the posterior distribution at step k can be expressed using the posterior distribution at step $k - 1$. In Section 6.4.2, we have defined the filtering problem at step k as computing the posterior distribution $p(x_k|y_{1:k}, u_{1:k})$. Applying the Bayes theorem to this posterior, we have

$$p\left(x_k \mid y_{1:k}, u_{1:k}\right) = \frac{p\left(y_k \mid x_k, y_{1:k-1}, u_{1:k}\right) p(x_k \mid y_{1:k-1}, u_{1:k})}{p(y_k \mid y_{1:k-1}, u_{1:k})}$$

$$= \eta_k p\left(y_k \mid x_k, y_{1:k-1}, u_{1:k}\right) p(x_k \mid y_{1:k-1}, u_{1:k}),$$

$$(6.19)$$

where η_k is the normalizer factor for step k. Due to the Markov assumption, the measurement y_k at step k is fully defined by the state x_k, i.e.,

$$p\left(y_k \mid x_k, y_{1:k-1}, u_{1:k}\right) = p(y_k \mid x_k). \tag{6.20}$$

This allows us to simplify Equation (6.19) as follows:

$$p\left(x_k \mid y_{1:k}, u_{1:k}\right) = \eta_k p(y_k \mid x_k) p(x_k \mid y_{1:k-1}, u_{1:k}). \tag{6.21}$$

To see what is involved in the term $p(x_k \mid y_{1:k-1}, u_{1:k})$ in the above equation, we expand it by integrating over x_{k-1} using the Chapman–Kolmogorov equation:

$$p\left(x_k \mid y_{1:k-1}, u_{1:k}\right)$$

$$= \int p\left(x_k \mid x_{k-1}, y_{1:k-1}, u_{1:k}\right) p(x_{k-1} \mid y_{1:k-1}, u_{1:k}) dx_{k-1}. \tag{6.22}$$

In the term $p\left(x_k \mid x_{k-1}, y_{1:k-1}, u_{1:k}\right)$ in Equation (6.22), based on the Markov assumption, the state x_{k-1} is a complete representation of step $k-1$ for predicting the future state x_k. Thus, we have

$$p(x_k \mid x_{k-1}, y_{1:k-1}, u_{1:k}) = p(x_k \mid x_{k-1}, u_k). \tag{6.23}$$

For the other term $p(x_{k-1} \mid y_{1:k-1}, u_{1:k})$ in Equation (6.22), since the external input u_k arrives after x_{k-1} and is assumed to be independent of the state, it can be safely omitted from that term. In other words,

$$p\left(x_{k-1} \mid y_{1:k-1}, u_{1:k}\right) = p(x_{k-1} \mid y_{1:k-1}, u_{1:k-1}). \tag{6.24}$$

Using Equations (6.23) and (6.24) to replace the corresponding terms in Equation (6.22), we have

$$p\left(x_k \mid y_{1:k-1}, u_{1:k}\right)$$

$$= \int p\left(x_k \mid x_{k-1}, u_k\right) p(x_{k-1} \mid y_{1:k-1}, u_{1:k-1}) dx_{k-1}. \tag{6.25}$$

Then, using Equation (6.25) to replace the term $p\left(x_k \mid y_{1:k-1}, u_{1:k}\right)$ in Equation (6.21), the posterior distribution is computed as follows:

$$p\left(x_k \mid y_{1:k}, u_{1:k}\right) = \eta_k \, p(y_k|x_k)$$

$$\int p\left(x_k \mid x_{k-1}, u_k\right) p(x_{k-1}|y_{1:k-1}, u_{1:k-1}) dx_{k-1}. \quad (6.26)$$

Equation (6.26) shows that the posterior distribution $p(x_k \mid y_{1:k}, u_{1:k})$ at step k can be expressed using three terms:

- $p(y_k|x_k)$, which is the measurement density (Equation (6.9)) defined by the measurement model;
- $p\left(x_k \mid x_{k-1}, u_k\right)$, which is the state transition density (Equation (6.8)) defined by the state transition model;
- $p(x_{k-1}|y_{1:k-1}u_{1:k-1})$, which is the posterior distribution from the previous step $k-1$.

This is a recursive formula because the posterior distribution at step k utilizes the posterior distribution from step $k-1$. This recursive formula leads to a general procedure for carrying out sequential Bayesian filtering (described in the following section). To make the recursion work, an initial belief of the state $p(x_0)$ is needed. Typically, $p(x_0)$ is assumed to be known or can be estimated based on some knowledge.

It is important to note that Equation (6.26) does not always have a closed-form solution for an arbitrary application because one cannot typically compute the integrals in the equation as well as the normalizer factor η_k in each step. An exception is the *linear Gaussian system* to be defined later, for which a closed-form solution exists, which serves as the foundation for the Kalman filter. For other general cases, approximation techniques are needed in order to compute the posterior distribution. In particular, the approximation based on the Monte Carlo method has been developed, which serves as the foundation for the particle filters.

6.5.2 *The prediction–update procedure*

From a functional point of view, the computation involved in Equation (6.26) can be divided into two sub-steps, named *Prediction* and *Update*. The corresponding computations for the two sub-steps are defined in Equations (6.25) and (6.21), respectively, and are restated as follows for convenience:

- Prediction:

$$p\left(x_k \mid y_{1:k-1}, u_{1:k}\right)$$

$$= \int p\left(x_k \mid x_{k-1}, u_k\right) p(x_{k-1}|y_{1:k-1}, u_{1:k-1})dx_{k-1}. \quad (6.27)$$

- Update:

$$p\left(x_k \mid y_{1:k}, u_{1:k}\right) = \eta_k p(y_k|x_k) p(x_k|y_{1:k-1}, u_{1:k}). \quad (6.28)$$

The outputs of these two sub-steps are the beliefs of x_k before and after incorporating the measurement data y_k, respectively. In the following description, we use $\overline{bel}(x_k)$ to denote the prior distribution $p\left(x_k \mid y_{1:k-1}, u_{1:k}\right)$, which is the output of the prediction sub-step. We use $bel(x_k)$ to denote the posterior distribution $p\left(x_k \mid y_{1:k}, u_{1:k}\right)$, which is the output of the update sub-step.

The two sub-steps form a *prediction–update procedure* used in each step (or each cycle) of sequential Bayesian filtering. Note that we use the term *sub-step* to indicate that the prediction and update are sub-routines of each step of sequential Bayesian filtering. When the context is clear, we will omit the prefix "sub-" and call them directly as the *prediction step* and *update step*.

Figure 6.8 shows the prediction–update procedure in one step of sequential Bayesian filtering. In the figure, $bel(x_{k-1})$ represents the posterior distribution $p(x_{k-1}|y_{1:k-1}, u_{1:k-1})$ computed from step $k - 1$. This is the input to the model-based prediction, which produces an output that is the prior distribution (denoted by $\overline{bel}(x_k)$) of step k. $\overline{bel}(x_k)$ then becomes the input of the measurement-based update, which produces an output that is the posterior distribution (denoted by $bel(x_k)$) of step k. $bel(x_k)$ stands for the estimation result of step k and also serves as the starting point for the next cycle of filtering. To explicitly show the role of the measurement data and to avoid confusion with the y_k variable used in the measurement model, Figure 6.8 uses the notation y_k^{md} to denote the actual measurement data at step k. Accordingly, the likelihood probability of observing y_k^{md} for a given x_k is denoted by $p(y_k = y_k^{md}|x_k)$.

More specifically, the prediction sub-step generates a prediction of the probability distribution of the state using the previous state belief $bel(x_{k-1})$ and the state transition model $p\left(x_k \mid x_{k-1}, u_k\right)$. The result

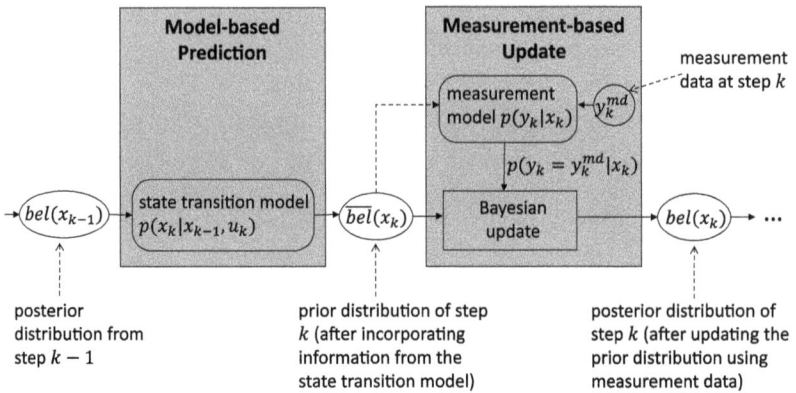

Figure 6.8.   The prediction–update procedure in sequential Bayesian filtering.

of the prediction step is the prior distribution $\overline{bel}(x_k)$ that represents the belief of the state after incorporating information from the state transition model (but not from the measurement data yet).

After the prediction step, the update sub-step updates the prior distribution using the measurement data y_k^{md}. This is a Bayesian update — it first computes the likelihood probabilities $p(y_k = y_k^{md}|x_k)$ for all x_k in the state space and then uses the likelihood probabilities to update the prior distribution. The computation of the likelihood probabilities are based on the measurement model $p(y_k|x_k)$ and the measurement data y_k^{md} (details in Section 6.6). The result of the update step is the posterior distribution $bel(x_k)$ that represents the belief of the state after incorporating information from both the state transition model and the measurement data.

The prediction-update procedure is invoked in each step of sequential Bayesian filtering. From a temporal point of view, the prediction sub-step evolves the belief of a state from the previous time step to the current time step, while the update sub-step updates the belief within the same time step. Figure 6.9 illustrates how belief is evolved and updated over multiple time steps. In the figure, a state's probability distribution is visualized by a dot and a circle (similar to the mean and covariance error ellipse for a Gaussian distribution). The bold lines between two consecutive time steps illustrate the state changes over time, and the star symbols at each time step illustrates the measurement data. As can be seen, each step predicts a new probability distribution from the previous step and then updates the probability distribution using the measurement data.

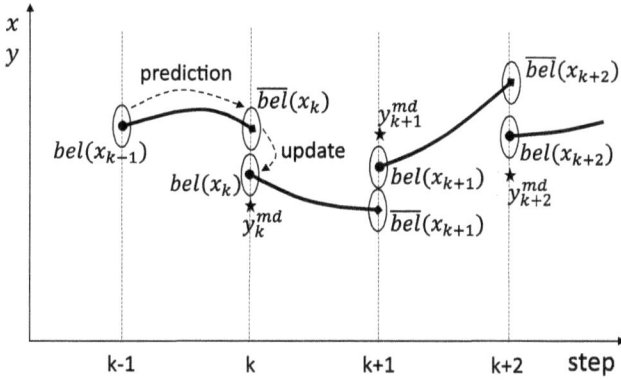

Figure 6.9. Illustration of the prediction–update procedure over time.

6.5.3 *The sequential Bayesian filtering algorithm*

The prediction–update procedure described above gives rise to a general sequential Bayesian filtering algorithm. Algorithm 6.1 shows this algorithm in a pseudo-algorithmic form.

Algorithm 6.1: The Sequential Bayesian Filtering Algorithm.

1. Initialization, $k = 0$:
   - Initialize $bel(x_0)$.
   - Advance the time step by setting $k = 1$.

2. Prediction (inputs: $(bel(x_{k-1}), u_k)$):
   - Predict the prior distribution using $bel(x_{k-1}), u_k$, and the state transition model: $\overline{bel}(x_k) = \int p(x_k|x_{k-1}, u_k) \, bel(x_{k-1}) dx_{k-1}$.

3. Update (inputs: $\overline{bel}(x_k), y_k^{md}$):
   - Compute the likelihood probabilities $p(y_k = y_k^{md}|x_k)$ for all x_k using the measurement data y_k^{md} and measurement model. See details in Section 6.6.
   - Compute the posterior distribution according to the Bayes theorem: $bel(x_k) = \eta_k p(y_k = y_k^{md}|x_k)\overline{bel}(x_k)$.

4. Advance the time step and start a new cycle:
   - Set $k \to k + 1$, and go to step 2.

The algorithm starts from an initialization step and then runs in a cyclic fashion. Each cycle corresponds to a time step in data assimilation and includes the prediction–update procedure. The prediction step (step 2) produces a predicted belief using the state transition model, starting from the posterior distribution of the previous step. Afterward, the update step (step 3) first computes the likelihood probabilities for all x_k using the measurement data and the measurement model (details in Section 6.6) and then updates the predicted belief according to the Bayes theorem. The update step produces a new posterior distribution that is the new state estimate after incorporating information from both the state transition model and measurement data. Then, the algorithm advances the time step and starts a new data assimilation cycle (step 4).

The general structure of sequential Bayesian filtering shown in Algorithm 6.1 is rarely realized in its original form. This is because the formulas that are involved in the prediction and update steps often do not have closed-form solutions, and thus, it is impossible to perform the belief update exactly. To address this problem, different methods have been developed to realize sequential Bayesian filtering that are applicable to different situations. The *Kalman filter* and *particle filter* are the two most well-known methods for realizing sequential Bayesian filtering. Sections 6.7 and 6.8 describe the Kalman filter and particle filter, respectively.

6.5.4 *Illustrative example*

This section provides an illustrative example of sequential Bayesian filtering. The example considers a mobile agent that moves on a one-dimensional road. The agent does not know its exact position. However, it carries a temperature sensor that measures the temperature of its location every 1 s. The temperatures are influenced by an outdoor fireplace that is located at a known position $x_{fireplace} = 80.0$. Due to the influence of the fireplace, the temperatures of different positions are different. The measurement data are noisy because of the precision constraint of the temperature sensor as well as disturbances from the environment. The agent moves toward right with a desired speed of 15.0 m/s (this information is known by the agent). However, the movement is not perfect due to the road condition and imperfect speed control.

In this example, the agent's state is its position, denoted by x. The measurement data are the temperature measurement every 1 s, denoted by y. The update of the agent's position (i.e., the state transition model) is described by a basic kinematic equation that has a constant moving speed and added Gaussian noise. The measurement model describes the temperature as a function of the position based on its distance to the fireplace: The closer to the fireplace, the higher the temperature (the detailed measurement model and the way of computing likelihood probabilities will be described in Section 6.6). The goal of sequential Bayesian filtering is to estimate the agent's dynamically changing position based on the temperature measurement data.

Figures 6.10(a)–(e) illustrate how sequential Bayesian filtering works in the first several time steps (each time step is 1 s). The top part of each figure shows the agent's true position (which is unknown to the agent), the location of the fireplace, and the temperature curve across the one-dimensional road. As can be seen, the temperature at the position of the fireplace is the highest ($40°$C); it gradually decreases for the positions that are away from the fireplace. In the following, we explain the figures in detail.

Figure 6.10(a) shows the initialization step (time 0) and the prediction step of time 1. In this example, the agent's true initial position is on the left side of the fireplace. We assume the agent starts with zero knowledge about where it is. In this case, the initial belief $bel(x_0)$ is that the agent could be anywhere on the road. This is illustrated by a visually flat probability distribution across the road. Starting from this initial belief, the prediction step updates the belief of the agent's new position (i.e., $\overline{bel}(x_1)$) at time 1. Since the initial belief is that the agent could be anywhere on the road, the predicted belief is still that the agent could be anywhere. This makes sense because it has not incorporated any information from measurement data yet.

Figure 6.10(b) shows the update step of time 1 using the temperature measurement data. It is assumed that the temperature measurement at this time step is $y_1^{md} = 35.3°$C, which is marked by a cross sign on the vertical axis of the top figure in Figure 6.10(b). The likelihood probabilities of observing this temperature for different positions of the road are shown in the middle figure in Figure 6.10(b). As can be seen, the likelihood is the highest for the two positions (marked as $L1$ and $L2$) that are on the left and right sides of

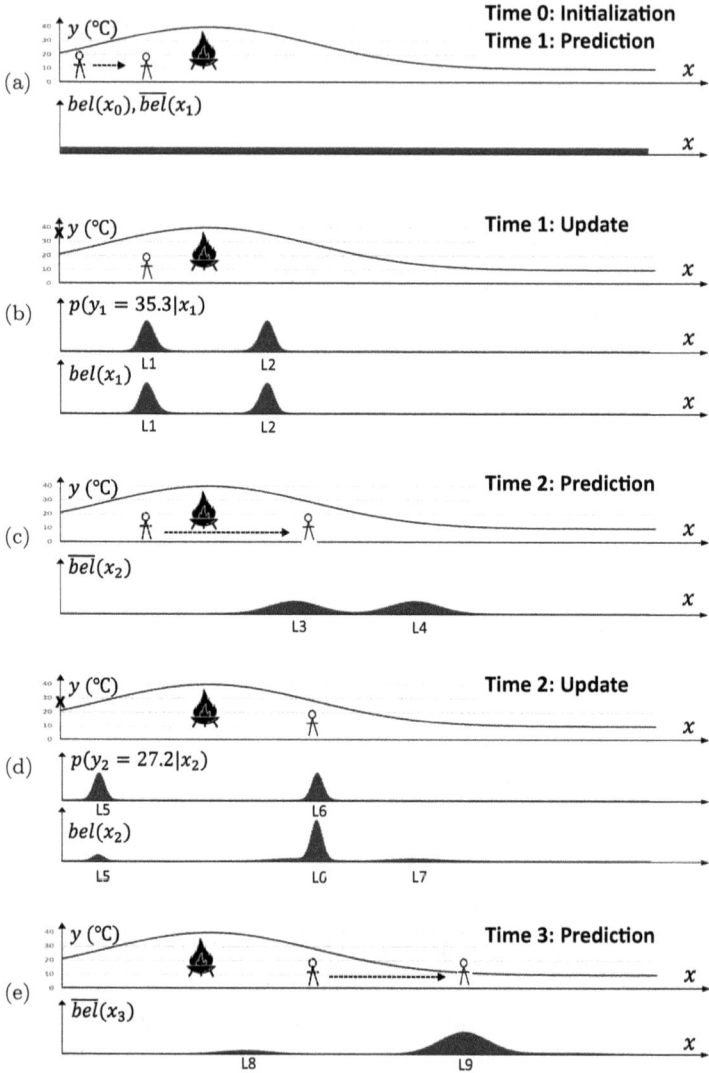

Figure 6.10.   Illustration of sequential Bayesian filtering.

the fireplace. The likelihoods decrease for other positions and become close to zero for the positions that are far away from the fireplace (because it is highly unlikely to have a temperature measurement of 35.3°C for those positions). $\overline{bel}(x_1)$ is then updated by multiplying the corresponding likelihood probabilities (see Equation (6.28)).

Since $\overline{bel}(x_1)$ is flat across the road, the resulting posterior distribution $bel(x_1)$ would have the same shape as the likelihood curve. This means the agent's position is most likely at or around $L1$ or $L2$, as indicated by the two "bumps" of the probability distribution of $bel(x_1)$. The width of the bump is an indication of the uncertainty of the estimate. At time 1, the positions $L1$ and $L2$ have the same probability values because the information from the measurement data cannot distinguish the two yet.

Figure 6.10(c) shows the prediction step of time 2, when the agent has moved to the right side of the fireplace. The prediction starts from $bel(x_1)$ and generates a belief $\overline{bel}(x_2)$ for the agent's new position based on the state transition model. As can be seen, compared to $bel(x_1)$, $\overline{bel}(x_2)$ has shifted to the right, with positions $L3$ and $L4$ having the highest probabilities. The bumps of $L3$ and $L4$ result from shifting the $L1$ and $L2$ bumps from $bel(x_1)$, respectively. The $L3$ and $L4$ bumps become lower and wider because the process noise associated with the movement adds new uncertainties to the belief of the state.

Figure 6.10(d) shows the update step of time 2. At this step, the temperature measurement is assumed to be $y_2^{md} = 27.2°C$. The likelihood probabilities of observing this temperature for different positions are shown in the middle figure in Figure 6.10(d). As can be seen, the likelihood probabilities concentrate at the positions $L5$ and $L6$. Updating $\overline{bel}(x_2)$ using these likelihood probabilities, we obtain the posterior distribution $bel(x_2)$ that is shown in the bottom figure in Figure 6.10(d). One can see that the belief is now dominated by the bump at $L6$, which is the result of multiplying the $L3$ bump in $\overline{bel}(x_2)$ and the $L6$ bump in the likelihood curve. Also, one can see that $L6$ is indeed close to the true position of the agent. The $L4$ bump of $\overline{bel}(x_2)$ is significantly diminished (it becomes the $L7$ bump of $bel(x_2)$) because the likelihood probabilities for that region are very small. $bel(x_2)$ also has a small bump at $L5$. For this region, the likelihood probabilities are high; however, the predicted belief $\overline{bel}(x_2)$ indicates that the probability for the agent to be in this region is very low. The small bump at $L5$ is the result of incorporating information from both the predicted state belief and the measurement data.

Figure 6.10(e) shows the prediction step of time 3, when the agent has further moved to the right. The prediction starts from $bel(x_2)$ and generates $\overline{bel}(x_3)$. As can be seen, the predicted belief shows that the agent is most likely at the position $L9$, which is the true position of the agent. There is a small possibility for the agent to be at the position $L8$ due to the shift of the $L5$ bump from $bel(x_2)$. The $L7$ bump of $bel(x_2)$ is almost eliminated due to the uncertainties added by the process noise.

Afterward, $\overline{bel}(x_3)$ will be updated by the measurement data at time 3 (not shown in the figure). This process will continue in a cyclic fashion as the agent moves and new measurement data arrive over time.

6.6 Likelihood Computation from Measurement Data

A key task of the update step of sequential Bayesian filtering is to compute the likelihood probabilities that are needed for updating the prior distribution. A likelihood probability defines the probability of observing y_k, which is the measurement data for a given state x_k. It is expressed as a conditional probability $p(y_k|x_k)$ that describes how likely it is to observe the data y_k assuming x_k is true. To explicitly show the role of measurement data, in the following description, we use the notation $p(y_k = y_k^{md}|x_k)$ to denote the likelihood probability of observing y_k^{md} for a given x_k. In this notation, y_k is a random variable and y_k^{md} is the actual measurement.

The concept of likelihood probability is often a source of confusion for beginners. The following example helps to illustrate this concept. Consider a mobile agent moving on a one-dimensional road, carrying a GPS sensor that returns noisy measurement data about the agent's position. In this example, the agent's state is its position (denoted by x_k) and the measurement data is the GPS measurement (denoted by y_k). Let us assume that at step k, the GPS measurement is 40.0, i.e., $y_k^{md} = 40.0$. The likelihood probabilities of observing this measurement data would be different if the agent is at different positions. For example, if the agent is at position $x_k = 40.2$, the likelihood probability for the GPS measurement to be 40.0 would be pretty high. This is because the difference between the GPS measurement and the agent's position is small (0.2 in this case). Considering the fact that the GPS measurement is noisy, we can say that it is very likely to observe the measurement data $y_k^{md} = 40.0$ if the agent is at position $x_k = 40.2$. Now, let us consider the likelihood probability of observing $y_k = 40.0$ if the agent is at position $x_k = 100.0$. The likelihood probability in this case would be low. This is because the agent's position is far away from the GPS measurement, and thus, it is unlikely to observe $y_k^{md} = 40.0$ if the agent is at position $x_k = 100.0$. In the form of conditional probability, the former likelihood probability is denoted by $p(y_k = 40.0 | x_k = 40.2)$ and the latter is denoted by $p(y_k = 40.0 | x_k = 100.0)$. In this example, we have $p(y_k = 40.0 | x_k = 40.2) > p(y_k = 40.0 | x_k = 100.0)$.

More generally, a likelihood probability of observing $y_k = 40.0$ exists for every position that the agent can be on the one-dimensional road (i.e., the state space of x_k). The likelihood probabilities for different positions would be different, which can be expressed as a function of x_k, denoted by $p(y_k = 40.0 | x_k)$.

The above example illustrates the concept of likelihood. However, it does not show how exactly to compute the probability values for different x_k. This is where the measurement model comes into play. A measurement model carries information about the mapping between a state vector and a measurement vector. It also captures the noises/uncertainties involved in the measurement. This information can be used to compute the likelihood probability values, as described in the following.

6.6.1 *Scalar Gaussian measurement*

This section describes how to compute the likelihood probabilities for a special type of measurement data called *scalar Gaussian measurement*, which is the main type of measurement data considered in this book. A scalar Gaussian measurement is a measurement data that is scalar (i.e., containing a numerical value) and has additive Gaussian noise (i.e., the measurement noise follows a Gaussian distribution). Typically, the Gaussian noise is a zero-mean Gaussian noise with some variance value. Scalar Gaussian measurement covers measurement data from a wide range of sensors that return numerical measures and have Gaussian (or Gaussian-like) noises. Examples include temperature measures from temperature sensors, position measures from GPS sensors, and distance measures from sonar sensors. It also covers cases where a sensor's raw measurement data may not be scalar and Gaussian, but the raw data are processed to produce numerical measures that have Gaussian-like noises. For example, a thermal camera may be used to monitor the spread of a wildfire. In this example, the raw data from the thermal camera are thermal images. These images are processed to produce position measures of fire fronts, the noises of which can be modeled by Gaussian noises. In this example, the resulting position measurement data are scalar Gaussian measurements.

A scalar Gaussian measurement can include a single measurement variable or, more generally, a measurement vector that has multiple measurement variables. We name the former case as *univariate scalar Gaussian measurement* and the latter case as *multivariate scalar Gaussian measurement*. For the latter case, each variable of the measurement vector needs to be a scalar Gaussian measurement.

6.6.1.1   *Univariate scalar Gaussian measurement*

We first consider the univariate scalar Gaussian measurement that includes a single measurement variable. For a univariate scalar Gaussian measurement, the measurement model Equation (6.6) can be decomposed into two components: a component describing the mapping from x_k to y_k, and another component describing the additive Gaussian noise. The new form of the measurement model is described as follows:

$$y_k = MF(x_k) + \varepsilon_k \quad \varepsilon_k \sim N(0, \sigma^2), \tag{6.29}$$

where x_k is the state vector, y_k is the univariate measurement, $MF()$ is the function specifying the mapping from x_k to y_k, and ε_k is a zero-mean Gaussian noise with a variance of σ^2. The function $MF()$ takes the state x_k as the input and generates a scalar output in the measurement domain. The output of $MF()$ is called the *predicted measurement*, denoted by $\tilde{y}_k$, which is the computed measurement purely based on the mapping from x_k assuming there is no measurement noise. Note that $MF()$ could be an arbitrary function (linear or nonlinear, in equation form or a computer program) describing the mapping from x_k to y_k. Also, note that the linear Gaussian measurement model used in the Kalman filter (to be described in Section 6.7) is a special case of scalar Gaussian measurement where the $MF()$ is a linear function.

For scalar Gaussian measurement whose measurement model is described by Equation (6.29), the likelihood probability $p(y_k = y_k^{md}|x_k)$ for different x_k can be computed using the following equation:

$$p\left(y_k = y_k^{md} \mid x_k\right) = \frac{1}{\sigma\sqrt{2\pi}} e^{-\frac{1}{2}\left(\frac{MF(x_k)-y_k^{md}}{\sigma}\right)^2},\qquad(6.30)$$

where y_k^{md} is the actual measurement data. This equation is straightforward to understand: To calculate the likelihood probability of observing y_k^{md} assuming x_k is true, we first compute the predicted measurement using the mapping function, i.e., $\widetilde{y}_k = MF(x_k)$. We then calculate the likelihood probability for the sensor to sense the predicted measurement $\widetilde{y}_k$ but returns a y_k^{md} value (i.e., the actual measurement data). Since the measurement noise is Gaussian, the likelihood probability is defined by the probability density function of the Gaussian noise based on the difference between the predicted measurement ($\widetilde{y}_k$) and the actual measurement (y_k^{md}).

Consider the mobile agent example described earlier. Let us assume the measurement noise of the GPS sensor has zero mean and a variance of σ_{gps}^2 ($\sigma_{gps} = 2.0$). Then, the measurement model of this example is $y_k = x_k + \varepsilon_k$, where ε_k is the noise factor $\varepsilon_k \sim N(0, \sigma_{gps}^2)$. In this example, the mapping function $MF()$ is very simple, i.e., $MF(x_k) = x_k$. Figure 6.11 (top) shows the mapping relationship between x_k and y_k. Let us assume at step k, the measurement data y_k^m is 40.0. Using Equation (6.30), the likelihood probability of observing this measurement for different x_k would be

$$p\left(y_k = 40.0 \mid x_k\right) = \frac{1}{\sigma_{gps}\sqrt{2\pi}} e^{-\frac{1}{2}\left(\frac{x_k-40.0}{\sigma_{gps}}\right)^2}.\qquad(6.31)$$

Based on Equation (6.31), Figure 6.11 (bottom) shows the likelihood probability as a function of x_k. As can be seen, the likelihood probabilities over x_k follow a Gaussian curve: The mean of the Gaussian curve is 40.0, and the variance is σ_{gps}^2. In other words, the likelihood probability of observing $y_k = 40.0$ for different x_k follows a Gaussian distribution $p(y_k = 40.0 \mid x_k) \sim N(40.0, \sigma_{gps}^2)$. This is intuitive to interpret: The probability of observing $y_k = 40.0$ is the highest when $x_k = 40.0$. Otherwise, the farther that x_k is away from 40.0, the less likely it is to observe $y_k = 40.0$.

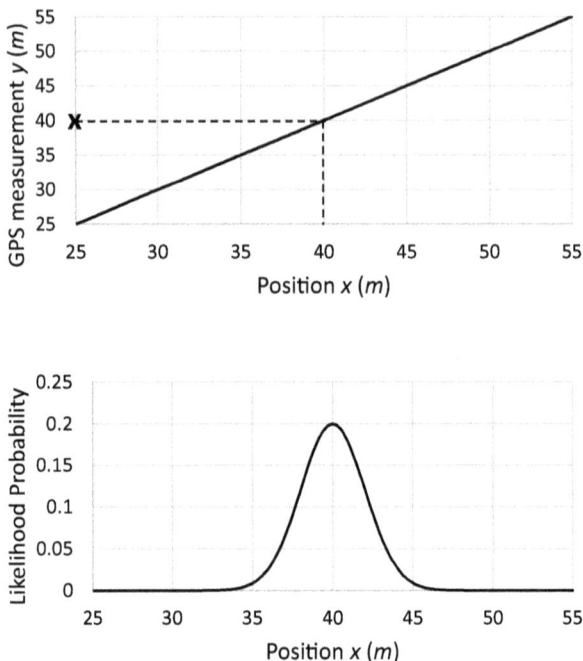

Figure 6.11. Illustration of likelihood probability for a linear mapping function $MF()$: (top) mapping between position x_k and measurement y_k. The real measurement data $y_k^{md} = 40.0$ is marked by the cross sign on the vertical axis; (bottom) likelihood probability $p(y_k = 40|x_k)$ as a function of x_k.

The right-hand side of Equation (6.30) shows that the likelihood probability follows a Gaussian curve from the predicted measurement $\tilde{y}_k = MF(x_k)$ point of view. When $MF()$ is a linear function, the likelihood probability would follow a Gaussian curve from the x_k point of view too. This is because a linear transformation does not modify the Gaussian property. This is the case for the example shown in Figure 6.11, whose $MF(x_k) = x_k$, which is a linear function. However, when $MF()$ is a nonlinear function, the mapping from x_k to y_k is nonlinear, and the likelihood probability will not follow a Gaussian curve from the x_k point of view.

Figure 6.12 illustrates an example of a nonlinear mapping from x_k to y_k and the associated likelihood probability. This example is based on the same scenario of a mobile agent moving on a one-dimensional road. Instead of using a GPS sensor, the agent carries a temperature sensor. The temperature sensor measures the temperature at the position of the agent. The temperatures are influenced by an outdoor fireplace that is located at a known position $x_{heat} = 80.0$.

We consider a simplified measurement model that describes the temperature as a function of the distance to the heat source: The temperature decreases according to a bell curve as the distance to the heat source increases. The measurement model that describes the mapping from x_k (position) to y_k (temperature measurement) is given as

$$y_k = T_c e^{\frac{-(x_k-80.0)^2}{\sigma_c^2}} + T_a + \varepsilon_k, \tag{6.32}$$

where $(x_k - 80.0)$ is the distance to the heat source, $T_c(°C)$ is the temperature rise above the ambient temperature of the heat source and is set to be $30°C$, $T_a(°C)$ is the ambient temperature and is set to be $10°C$, σ_c is a constant and is set to 15 m, and ε_k is the measurement noise and is assumed to be a zero-mean Gaussian noise $\varepsilon_k \sim N(0, \sigma_{temp}^2)$, where $\sigma_{temp} = 2.0$. In this example, the mapping function $MF(x_k) = T_c e^{\frac{-(x_k-80.0)^2}{\sigma_c^2}} + T_a$. This is a nonlinear function, which is displayed in Figure 6.12 (top). As can be seen, the temperature at the position of the heat source is the highest ($40°C$). It decreases as the distance to the heat source increases according to a bell curve.

Let us assume at step k, the measurement data $y_k^{md} = 30.0°C$. Using Equation (6.30) and listing the term $MF()$ as a separate equation, the likelihood probabilities $p(y_k = 30.0|x_k)$ of observing this measurement for different x_k can be calculated as follows:

$$p(y_k = 30.0 \mid x_k) = \frac{1}{\sigma_{temp}\sqrt{2\pi}} e^{-\frac{1}{2}\left(\frac{MF(x_k)-30.0}{\sigma_{temp}}\right)^2},$$

$$MF(x_k) = T_c e^{\frac{-(x_k-80.0)^2}{\sigma_c^2}} + T_a. \tag{6.33}$$

Figure 6.12 (bottom) shows the likelihood probabilities $p(y_k = 30.0|x_k)$ as a function of x_k. As can be seen, the probability of observing $y_k^{md} = 30.0°C$ is the highest for the positions $x_L \approx 70.4$ and $x_R \approx 89.6$, which are indicated in the figure. These two positions are located on the left and right sides of the heat source, respectively,

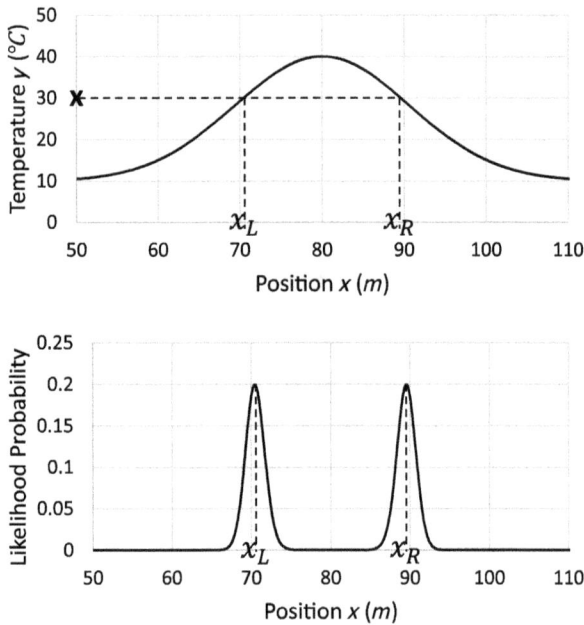

Figure 6.12.   Illustration of the likelihood probability for a nonlinear mapping function $MF()$: (top) mapping between position x_k and measurement y_k. The real measurement data $y_k^{md} = 30.0°C$ is marked by the cross sign on the vertical axis; (bottom) likelihood probability $p(y_k = 30.0|x_k)$ as a function of x_k.

which have the predicted measurement of exactly 30.0. The probability values decrease for the positions that are away from x_L and x_R: The further the distances, the smaller the probability values. It is clear that the curve of the likelihood probabilities in this example is more complex than that in the previous example. The complexity stems from the fact that the measurement model is a nonlinear model, even though the measurement noise is still Gaussian.

6.6.1.2   *Multivariate scalar Gaussian measurement*

More often, the measurement y_k is a vector, carrying measurement data from multiple sensors. In these cases, $y_k = [y_{1,k}, y_{2,k}, \cdots, y_{m,k}]^T$, where m is the dimension of the vector and $y_{i,k}$ is the ith element of the measurement vector. Similar to that for the univariate scalar Gaussian measurement, the measurement model for the multivariate scalar Gaussian measurement can be decomposed into two components: a component of the mapping function and a component of the Gaussian noise, as follows:

$$y_k = MF(x_k) + \varepsilon_k \quad \varepsilon_k \sim N(0, \Sigma), \tag{6.34}$$

where $MF()$ is the mapping function from state vector x_k to measurement vector y_k, ε_k is a Gaussian noise vector that is of the same dimension as the measurement vector. The distribution of ε_k is a multivariate Gaussian with zero mean and a covariance of Σ.

The same approach to computing the likelihood probability for the univariate scalar Gaussian measurement can be extended to the multivariate scalar Gaussian measurement. Specifically, for the multivariate scalar Gaussian measurement whose measurement model is descried by Equation (6.34), the likelihood probability $p(y_k = y_k^{md}|x_k)$ can be computed using the following equation:

$$p\left(y_k = y_k^{md} \mid x_k\right) = \frac{\exp\left(-\frac{1}{2}(MF(x_k) - y_k^{md})^T \Sigma^{-1} (MF(x_k) - y_k^{md})\right)}{\sqrt{(2\pi)^m |\Sigma|}}, \tag{6.35}$$

where y_k^{md} is the vector of the actual measurement data, Σ is the covariance matrix, Σ^{-1} is the inverse of the covariance matrix, and $|\Sigma|$ is the determinant of Σ.

Typically, sensor data are assumed to be independent of each other. Thus, each sensor measurement can be treated as a univariate scalar Gaussian measurement that has its own Gaussian noise. With the independence assumption, the likelihood probability $p(y_k = y_k^{md}|x_k)$ can be simplified as the product of the individual measurement likelihoods, i.e.,

$$p\left(y_k = y_k^{md} \mid x_k\right) = \prod_{i=1}^{m} p\left(y_{i,k} = y_{i,k}^{md} \mid x_k\right), \tag{6.36}$$

where $y_{i,k}$ is the ith element of the measurement vector and $y_{i,k}^{md}$ is the actual measurement data for the ith element. The measurement likelihood for the ith element, i.e., $p\left(y_{i,k} = y_{i,k}^{md} \mid x_k\right)$, is computed using Equation (6.30) by treating the ith element as a univariant scalar Gaussian measurement.

6.6.2 *Non-scalar Gaussian measurement*

While the scalar Gaussian measurement covers measurement data from a wide range of sensors, there are measurement data that do not satisfy the scalar and Gaussian requirements. This can happen in a number of ways. One way is that the measurement noise (or measurement error) of the data is not Gaussian. Consider the range finder sensors (e.g., laser range finder sensors) studied by Thrun *et al.* (2005) as an example. One of the measurement errors of these sensors is that they sometimes fail to detect obstacles altogether, resulting in a so-called *max-range measurement*: The sensor returns its maximum allowable value. The measurement errors due to these failures are not Gaussian, as it is not meaningful to add a Gaussian noise to the returned maximum allowable value to model the possibility of a failure. In this example, the measurement data are scalar (i.e., the maximum allowable range value), but the noises are not Gaussian, and thus, the measurement data is not a scalar Gaussian measurement. Another way is that the measurement data are not scalar. For example, a type of measurement data that often exist in discrete event applications is the *event-based measurement data*, e.g., if a job is finished or not. These event-based measurements are triggered only when some conditions are satisfied, and they have categorical values (e.g., *stage1₋ finished, stage2₋ finished, completed*), as opposed to numerical values. These measurement data are not scalar Gaussian measurement.

For non-scalar Gaussian measurement, the computation of likelihoods cannot use the equations presented in the previous sections. Instead, one needs to examine the properties of the measurement model, including the state–measurement mapping relationship and the noise characteristics, to compute the likelihood probabilities accordingly. Sometimes, the likelihood probabilities may be approximated in an application-specific way. For example, the range finder sensor failure described above, which is used in robotic applications, can be modeled by a point-mass distribution centered at the maximum allowable value (Thrun *et al.*, 2005). For the event-based measurement data, the measurement models typically include some kind of quantization mechanism. Examples of how to handle quantized measurements for event-based state estimation can be found in the works of Sijs and Lazar (2012) and Davar and Mohammadi (2017). This book focuses only on scalar Gaussian measurement.

6.7 Kalman Filter

The Kalman filter was invented by Rudolf E. Kalman in the 1950s as a technique for filtering and prediction for linear systems. It is one of the most widely used data assimilation methods and is the foundation for several extensions (such as the extended Kalman filter and ensemble Kalman filter) that are used for data assimilation. The Kalman filter uses the Gaussian representation to represent beliefs. It thus works only for continuous states.

6.7.1 *Linear Gaussian state-space model*

The Kalman filter deals with the so-called *linear Gaussian systems* whose state transition model and measurement model must be linear and the noises that are involved must be Gaussian noises. Mathematically, the state transition model and the measurement model are specified by

$$x_k = A_k x_{k-1} + B_k u_k + \gamma_k, \tag{6.37}$$

$$y_k = C_k x_k + \varepsilon_k, \tag{6.38}$$

where x_{k-1} and x_k are the state vectors at times $k-1$ and k, respectively, u_k is the input vector, and y_k is the measurement vector at time k. Here, x_k, u_k, and y_k are all vertical vectors and may have different dimensions. A_k, B_k, and C_k are matrices. A_k is the *transition matrix* with a size of $n \times n$, where n is the dimension of the state vector x_k. B_k is the *control matrix* with a size of $n \times l$, with l being the dimension of the input vector u_k. C_k is the *observation matrix* with a size of $m \times n$, with m being the dimension of the measurement vector y_k.

γ_k is the process noise and must be a Gaussian random vector having the same dimension as that of x_k. Its mean is zero and its covariance is an $n \times n$ matrix denoted by Q_k, i.e, $\gamma_k \sim N(0, Q_k)$. ε_k is the measurement noise and must be a Gaussian random vector too.

Its dimension is the same as that of y_k. Its mean is zero and its covariance is an $m \times m$ matrix denoted by R_k, i.e., $\varepsilon_k \sim N(0, R_k)$. The purpose of the process noise is to model the uncertainty involved in the state transition, and the purpose of the measurement noise is to model the uncertainty involved in the measurement.

Equations (6.37) and (6.38) are special cases of Equations (6.5) and (6.6), respectively. The fact that Equation (6.37) is described in a matrix form means that x_k is a linear transformation from x_{k-1} and u_k plus the Gaussian noise γ_k. Similarly, y_k is a linear transformation from x_k plus the Gaussian noise ε_k. We name these models that have linear transformations and Gaussian noises *linear Gaussian models*.

The linear Gaussian models have important consequences for the computation of the prior and posterior distributions of the state vector. Specifically, if x_{k-1} has a Gaussian distribution with a mean of μ_{k-1} and a covariance of Σ_{k-1}, after a state transition using the linear Gaussian model defined in Equation (6.37), the prior distribution of the predicted state would still be Gaussian. In this case, it is possible to precisely derive the solution for the prediction step (Equation (6.27)) of sequential Bayesian filtering to compute the prior distribution $\overline{bel}(x_k)$. Specifically, $\overline{bel}(x_k)$ would be a Gaussian distribution, i.e., $\overline{bel}(x_k) \sim N(\overline{\mu}_k, \overline{\Sigma}_k)$, whose mean $\overline{\mu}_k$ and covariance $\overline{\Sigma}_k$ have the following respective values:

$$\overline{\mu}_k = A_k \mu_{k-1} + B_k u_k, \tag{6.39}$$

$$\overline{\Sigma}_k = A_k \Sigma_{k-1} A_k^T + Q_k. \tag{6.40}$$

Similarly, when working with a linear Gaussian measurement model as defined in Equation (6.38), if the prior distribution is Gaussian, then the posterior distribution resulting from the update step (Equation (6.28)) is also Gaussian. One can precisely derive the solution for the Gaussian posterior distribution. Let $\overline{bel}(x_k)$ be the prior distribution that has a mean of $\overline{\mu}_k$ and a covariance of $\overline{\Sigma}_k$ and y_k^{md} be the measurement data at step k. Let $bel(x_k)$ be the resulting posterior distribution from the update step. Then, $bel(x_k)$ is a Gaussian distribution whose mean μ_k and covariance Σ_k have the following respective values:

$$\mu_k = \overline{\mu}_k + K_k(y_k^{md} - C_k \overline{\mu}_k), \tag{6.41}$$

$$\Sigma_k = (I - K_k C_k)\overline{\Sigma}_k, \tag{6.42}$$

where K_k is called the *Kalman gain*, which is an $n \times m$ matrix and defined as follows. In the equation, the superscript T marks the transpose of a matrix and the superscript -1 marks the inverse of a matrix:

$$K_k = \overline{\Sigma}_k C_k^T (C_k \overline{\Sigma}_k C_k^T + R_k)^{-1}. \tag{6.43}$$

Detailed derivations of the above solutions are omitted in this book. Interested readers may refer to the work of Thrun *et al.* (2005) for a step-by-step derivation of these equations.

The term $(y_k^{md} - C_k \overline{\mu}_k)$ in Equation (6.41) is often called the *innovation* or *measurement residual*, which specifies the difference between the actual measurement and the mean predicted measurement (computed as the product of the observation matrix and the mean predicted state). The innovation is a measure of the new information provided by the measurement data.

The Kalman gain K_k defines how the innovation $(y_k^{md} - C_k \overline{\mu}_k)$ is incorporated into the state estimate. It is computed based on Equation (6.43). Note that in Equation (6.43), the terms $\overline{\Sigma}_k C_k^T$ and $C_k \overline{\Sigma}_k C_k^T$ are computed from the covariance $(\overline{\Sigma}_k)$ of the predicted state belief, and the term R_k is the covariance of the measurement noise. Since covariances are directly related to uncertainties, an intuitive way of understanding the Kalman gain is to treat it as a relative weight defined by the uncertainties of the predicted state and the measurement data, as shown in the following. This relative weight determines the relative level of trust one has in the predicted state and the measurement data when combining information from the two sources:

$$K_k = \frac{\overline{\Sigma}_k C_k^T}{C_k \overline{\Sigma}_k C_k^T + R_k}$$

$$= \frac{\text{uncertainty of predicted state}}{\text{uncertainty of predicted state} + \text{uncertainty of measuerment data}}. \tag{6.44}$$

To better illustrate the above idea, we consider two extremes by computing the limits of $R_k \to 0$ and $\overline{\Sigma}_k \to 0$:

$$\lim_{R_k \to 0} \frac{\overline{\Sigma}_k C_k^T}{C_k \overline{\Sigma}_k C_k^T + R_k} = C_k^{-1}, \tag{6.45}$$

$$\lim_{\overline{\Sigma}_k \to 0} \frac{\overline{\Sigma}_k C_k^T}{C_k \overline{\Sigma}_k C_k^T + R_k} = 0. \tag{6.46}$$

Substituting the first limit into Equation (6.41), one can see that when R_k approaches 0, then μ_k would approach $C_k^{-1} y_k^{md}$. This means when the covariance R_k is small (meaning that the uncertainty of the measurement data is small), μ_k of $bel(x_k)$ is determined mostly by the measurement data y_k^{md}. On the other hand, substituting the second limit into Equation (6.41), one can see that when $\overline{\Sigma}_k$ approaches 0, then μ_k would approach $\overline{\mu}_k$. This means when the uncertainty of $\overline{bel}(x_k)$ is small (meaning that the predicted state has low uncertainty), μ_k of $bel(x_k)$ is determined mostly by the predicted state. Between these two extremes, the estimated state uses information from both the predicted state and measurement data according to the relative weight of the uncertainties of the two sources of information, as defined by the Kalman gain.

The Kalman filter is based on the solutions described in Equations (6.39)–(6.43). The fact that the Kalman filter computes precise solutions for the posterior distribution means that there is no approximation involved in the Kalman filter computation.

6.7.2   *The Kalman filter algorithm*

Algorithm 6.2 shows the Kalman filter algorithm, which follows the general structure of the sequential Bayesian filtering algorithm presented earlier. The algorithm is based on the linear Gaussian system described by Equations (6.37) and (6.38). y_k^{md} denotes the measurement data at step k.

Algorithm 6.2:. The Kalman Filter Algorithm.

1. Initialization, $k = 0$:

   - Initialize $bel(x_0)$ as a Gaussian distribution: $bel(x_0) \sim N(\mu_0, \Sigma_0)$.
   - Advance the time step by setting $k = 1$.

2. Prediction (input: $\mu_{k-1}, \Sigma_{k-1}, u_k$):

   - Compute the mean and covariance for the prior distribution $\overline{bel}(x_k) \sim (\overline{\mu}_k, \overline{\Sigma}_k)$:
     - $\overline{\mu}_k = A_k \mu_{k-1} + B_k u_k,$
     - $\overline{\Sigma}_k = A_k \Sigma_{k-1} A_k^T + Q_k.$

---

3. Update (input: $\overline{\mu}_k, \overline{\Sigma}_k, y_k^{md}$):
   - Compute the mean and covariance for the posterior distribution $bel(x_k) \sim (\mu_k, \Sigma_k)$:
     - $K_k = \overline{\Sigma}_k C_k^T (C_k \overline{\Sigma}_k C_k^T + R_k)^{-1}$,
     - $\mu_k = \overline{\mu}_k + K_k(y_k^{md} - C_k\overline{\mu}_k)$,
     - $\Sigma_k = (I - K_k C_k)\overline{\Sigma}_k$.
4. Advance the time step and start a new cycle:
   - Set $k \to k + 1$, and go to step 2.

In the Kalman filter, all the beliefs are Gaussian distributions characterized by a mean and a covariance. The initialization step is to provide an initial belief $bel(x_0)$ that is specified by $N(\mu_0, \Sigma_0)$. The mean μ_0 and covariance Σ_0 can be set based on the prior knowledge about a system's initial state. When there is a lack of such knowledge, a common practice is to assign the variances with very large values so that the Gaussian distribution spreads out to cover a large state space.

The prediction step takes $bel(x_{k-1}) \sim N(\mu_{k-1}, \Sigma_{k-1})$ and u_k as inputs and computes the mean $\overline{\mu}_k$ and covariance $\overline{\Sigma}_k$ for the prior distribution $\overline{bel}(x_k)$. The mean $\overline{\mu}_k$ and covariance $\overline{\Sigma}_k$ are computed using Equations (6.39) and (6.40).

The update step takes $\overline{bel}(x_k) \sim N(\overline{\mu}_k, \overline{\Sigma}_k)$ and y_k^{md} as inputs and computes the mean μ_k and covariance Σ_k for the posterior distribution $bel(x_k)$. To do this, it first computes the Kalman gain K_k according to Equation (6.43) and then uses K_k to compute the mean μ_k and covariance Σ_k based on Equations (6.41) and (6.42).

After the update step, the algorithm advances the time step and starts a new cycle.

The computations involved in the Kalman filter algorithm are mainly matrix manipulation and transformation, which can be very efficient even for high-dimensional state and measurement vectors. Since the state transition model is a linear Gaussian model, the mean and covariance of the belief from the state transition can be computed directly using the corresponding equations. There is no simulation involved in the computation.

6.7.3   *Extensions of the Kalman filter*

The Kalman filter is based on the assumptions that the state transition model and the measurement model are linear models and that the noises are Gaussian noises. Several extensions have been developed to expand the applicability of the Kalman filter approach. This section introduces the two most important extensions: *extended Kalman filter* (EKF) and *ensemble Kalman filter* (EnKF).

Both the EKF and EnKF address a major limitation of the Kalman filter: the linearity assumption of the state transition model and measurement model. This means EKF and EnKF work with nonlinear state transition and measurement models. Nevertheless, they still assume that the noises are Gaussian noises. The nonlinear state transition and measurement models mean that the resulting beliefs are no longer Gaussian, so the Kalman filter equations cannot be directly applied. The EKF and EnKF employ different approximations to work with the nonlinear models.

Without the linearity assumption, the state-space model can be written as

$$x_k = f\left(x_{k-1}, u_k\right) + \gamma_k, \tag{6.47}$$

$$y_k = g\left(x_k\right) + \varepsilon_k, \tag{6.48}$$

where x_k, y_k, and u_k are the state vector, measurement vector, and input vector, respectively, as described earlier. $\gamma_k \sim N(0, Q_k)$ and $\varepsilon_k \sim N(0, R_k)$ are the process noise and measurement noise, respectively, which are still zero-mean Gaussian noises. The nonlinear functions $f()$ and $g()$ define the state transition model and measurement model, respectively.

6.7.3.1   *Extended Kalman filter*

The key idea of the EKF is to linearize the state transition and measurement models using their partial derivatives with respect to the state value at the time of linearization so that the same equations from the Kalman filter can be used to update the mean and covariance of the state beliefs. The EKF still use the Gaussian representation to approximate beliefs. The mean μ_k and covariance Σ_k of the posterior distribution $bel(x_k)$ are computed as follows:

- Prediction:

$$\overline{\mu}_k = f(\mu_{k-1}, u_k), \tag{6.49}$$

$$\overline{\Sigma}_k = F_k \Sigma_{k-1} F_k^T + Q_k; \tag{6.50}$$

- Update:

$$K_k = \overline{\Sigma}_k G_k^T (G_k \overline{\Sigma}_k G_k^T + R_k)^{-1}, \tag{6.51}$$

$$\mu_k = \overline{\mu}_k + K_k (y_k^{md} - g(\overline{\mu}_k)), \tag{6.52}$$

$$\Sigma_k = (I - K_k G_k) \overline{\Sigma}_k, \tag{6.53}$$

Where μ_{k-1} and Σ_{k-1} are the mean and covariance of $bel(x_{k-1})$, respectively, computed from the previous step and y_k^{md} is the measurement data. F_k is the transition matrix and is defined as the Jacobian of the $f()$ function using the values of μ_{k-1} and u_k. Similarly, G_k is the observation matrix and is defined as the Jacobian of the $g()$ function using the value of $\overline{\mu}_k$. Specifically,

$$F_k = \frac{\partial f}{\partial x}\bigg|_{\mu_{k-1}, u_k}, \quad G_k = \frac{\partial f}{\partial x}\bigg|_{\overline{\mu}_k}. \tag{6.54}$$

As can be seen, the EKF equations (Equations (6.49)–(6.53)) follow the same structures as those in the Kalman filter (Equations (6.39)–(6.43)). The two sets of equations differ in the following ways: (1) Equation (6.49) computes the mean $\overline{\mu}_k$ using the $f()$ function as opposed to the matrix transformation as in the Kalman filter; (2) Equation (6.52) computes the predicted measurement using the $g()$ function as opposed to the matrix transformation as in the Kalman filter; (3) Equations (6.50), (6.51), and (6.53) use the Jacobians F_k and G_k instead of the corresponding linear system matrices A_k, B_k, and C_k as in the Kalman filter.

6.7.3.2 *Ensemble Kalman filter*

The key idea of the EnKF (Evensen, 1994; Houtekamer and Mitchell, 1998) is to approximate the state belief using a set of samples that is called an *ensemble*. This sample-based representation is a major deviation from the Gaussian representation used by the Kalman filter and the EKF. From the belief representation point of view, EnKF is similar to the particle filter (PF) that will be described later, although the two use different approaches to update the posterior distribution based on measurement data.

The sample-based representation of the EnKF means that the prediction step no longer computes the mean and covariance of a Gaussian distribution. Instead, it computes a new state for each sample. Similarly, the update step is not to calculate the mean and covariance of the posterior distribution. Instead, it updates each sample's value to generate a new set of samples representing the posterior distribution. Despite the above differences, EnKF still employs the same principle as that of the Kalman filter to carry out the update: It shifts the value of each sample using the Kalman gain and innovation, i.e., the difference between the actual measurement and the predicted measurement.

Let $\{x_{k-1}^{(1)}, x_{k-1}^{(2)}, \ldots, x_{k-1}^{(N)}\}$ be the ensemble of samples representing the posterior distribution from step $k-1$, where N is the size of the ensemble. Each sample is a concrete instantiation of the state. To obtain samples for the prior distribution (i.e., $\overline{bel}(x_k)$), the prediction step applies the state transition model (Equation (6.47)) to each sample in the ensemble. The resulting ensemble (called the *prior ensemble*) includes a set of new samples $\{\overline{x}_k^{(i)}; i = 1, \ldots, N\}$, where

$$\overline{x}_k^{(i)} = f\left(x_{k-1}^{(i)}, u_k\right) + \gamma_k^{(i)}, \quad \gamma_k^{(i)} \sim N(0, Q_k). \tag{6.55}$$

The update step is to update each $\overline{x}_k^{(i)}$ based on the measurement data y_k^{md}. The updated samples then represent the posterior distribution $bel(x_k)$. In the EnKF, the update for the ith sample is to shift the predicted state $\overline{x}_k^{(i)}$ using the Kalman gain (denoted by K_k) and the innovation:

$$x_k^{(i)} = \overline{x}_k^{(i)} + K_k(y_k^{md} - \tilde{y}_k^{(i)}), \tag{6.56}$$

$$\tilde{y}_k^{(i)} = g\left(\overline{x}_k^{(i)}\right) + \varepsilon_k^{(i)}, \quad \varepsilon_k^{(i)} \sim N(0, R_k), \tag{6.57}$$

$$K_k = \overline{\Sigma}_k G_k^T (G_k \overline{\Sigma}_k G_k^T + R_k)^{-1}, \tag{6.58}$$

where $\tilde{y}_k^{(i)}$ is the predicted measurement for the ith sample and is computed using the measurement model (Equation (6.48)). In Equation (6.58), the $\bar{\Sigma}_k$ is the covariance of the predicted state belief $\overline{bel}(x_k)$ and G_k is the observation matrix.

A key task of the EnKF is to compute the Kalman gain K_k. Equation (6.58) shows that K_k can be computed from the term $\bar{\Sigma}_k G_k^T$, the term $G_k \bar{\Sigma}_k G_k^T$, and the covariance of the measurement noise R_k. For general problems, the exact values of the first two terms ($\bar{\Sigma}_k G_k^T$ and $G_k \bar{\Sigma}_k G_k^T$) cannot be analytically derived. In the EnKF, a common approach is to approximate $\bar{\Sigma}_k$ based on the spread of the samples in the prior ensemble and to compute $\bar{\Sigma}_k G_k^T$ and $G_k \bar{\Sigma}_k G_k^T$ directly from the samples without explicitly specifying the observation matrix G_k (Houtekamer and Mitchell, 2001). Specifically, $\bar{\Sigma}_k G_k^T$ and $G_k \bar{\Sigma}_k G_k^T$ are computed from the samples in the following way, where $\widehat{\bar{x}_k}$ and $\widehat{g(\bar{x}_k)}$ represent the mean values computed using Equations (6.61) and (6.62), respectively:

$$\bar{\Sigma}_k G_k^T \equiv \frac{1}{N-1} \sum_{i=1}^{N} (\bar{x}_k^{(i)} - \widehat{\bar{x}_k})(g(\bar{x}_k^{(i)}) - \widehat{g(\bar{x}_k)})^T, \tag{6.59}$$

$$G_k \bar{\Sigma}_k G_k^T \equiv \frac{1}{N-1} \sum_{i=1}^{N} (g(\bar{x}_k^{(i)}) - \widehat{g(\bar{x}_k)})(g(\bar{x}_k^{(i)}) - \widehat{g(\bar{x}_k)})^T, \tag{6.60}$$

$$\widehat{\bar{x}_k} = \sum_{i=1}^{N} \bar{x}_k^{(i)}/N, \tag{6.61}$$

$$\widehat{g(\bar{x}_k)} = \sum_{i=1}^{N} g(\bar{x}_k^{(i)})/N. \tag{6.62}$$

Even though the EnKF represents belief distributions using ensemble samples, the implicit assumption it relies on is that the likelihood and prior are both Gaussian. Based on this assumption, in Equation (6.56), the samples are updated by shifting their values based on the Kalman gain and the innovation. Updating the samples by shifting their values allows the EnKF to avoid the degeneracy problem that exists in particle filters. As a result, the EnKF is known to converge to the true posterior for high-dimensional states even when using a relatively small ensemble size.

6.7.4   *Illustrative example*

To illustrate how the Kalman filter works, we consider a simple example that includes a mobile agent moving on a one-dimensional road with an unknown speed. Due to the imperfect speed control, the actual speed of the agent may vary over time. The precise positions and the speeds of the agent are unknown and need to be estimated. The agent is equipped with a GPS sensor that returns noisy measurement data about the agent's position every 1 s. The goal of data assimilation is to estimate the agent's dynamically changing position and its moving speed based on the noisy measurement data from the GPS sensor.

In this example, the state vector of the agent is $x = [x^p, x^v]^T$, where x^p stands for its position and x^v stands for its speed. A positive x^v means the agent moves to the right, and a negative x^v means the agent moves to the left. The measurement y is the GPS data about the agent's position. The state estimation is carried out every 1 s ($\Delta t = 1$) when a new measurement data point becomes available.

The uncertainty in the agent's speed is modeled as a zero-mean Gaussian noise γ^v. Due to road conditions and other environmental factors, the update of the agent's position involves uncertainty too, which is modeled as a zero-mean Gaussian noise γ^p. With these two noise terms, the state transition model of the agent is defined as follows:

$$x_k^p = x_{k-1}^p + \Delta t x_{k-1}^v + \gamma^p,$$
$$x_k^v = x_{k-1}^v + \gamma^v, \tag{6.63}$$

where x_{k-1}^p, x_k^p, x_{k-1}^v, and x_k^v are the agent's positions and speeds at time steps $k - 1$ and k, respectively, and Δt is the time interval between two steps and $\Delta t = 1$. γ^p is the position noise, and γ^v is the speed noise. In this example, $\gamma^p \sim N\left(0, 2.0^2\right)$ and $\gamma^v \sim N\left(0, 0.5^2\right)$. γ^p and γ^v are assumed to be independent of each other.

Using the state vector $x = \begin{bmatrix} x^p \\ x^v \end{bmatrix}$ and defining the noise vector $\gamma = \begin{bmatrix} \gamma^p \\ \gamma^v \end{bmatrix}$, we can rewrite Equation (6.63) in the matrix form:

$$x_k = \begin{bmatrix} 1 & \Delta t \\ 0 & 1 \end{bmatrix} x_{k-1} + \begin{bmatrix} \gamma^p \\ \gamma^v \end{bmatrix} = A x_{k-1} + \gamma, \tag{6.64}$$

where A is the transition matrix and γ is the process noise: $\gamma \sim N(\mu, Q)$, with $\mu = \begin{bmatrix} 0 \\ 0 \end{bmatrix}$ and $Q = \begin{bmatrix} 2.0^2 & 0 \\ 0 & 0.5^2 \end{bmatrix}$. Comparing Equation (6.64) with the state transition model defined in Equation (6.37), one can see that there is no Bu term here. This is because there is no external (control) input in this example. Furthermore, since A and γ are independent of the time step, the step index subscripts for A and γ are omitted.

The measurement noise is assumed to be a zero-mean Gaussian noise: $\varepsilon \sim N(0, 5.0^2)$. Let y_k be the measurement data at step k. In matrix form, we have

$$y_k = \begin{bmatrix} 1 & 0 \end{bmatrix} \begin{bmatrix} x_k^p \\ x_k^v \end{bmatrix} + \varepsilon = Cx_k + \varepsilon. \tag{6.65}$$

Based on the state transition model and measurement model (Equations (6.64) and (6.65)) described above and using $\Delta t = 1$, the matrices that define the linear Gaussian model of this system are

$$A = \begin{bmatrix} 1 & 1 \\ 0 & 1 \end{bmatrix}, B = \begin{bmatrix} 0 \\ 0 \end{bmatrix}, C = \begin{bmatrix} 1 & 0 \end{bmatrix}, Q = \begin{bmatrix} 2.0^2 & 0 \\ 0 & 0.5^2 \end{bmatrix}, \text{and } R = 5.0^2.$$

To run the Kalman filter algorithm, an initial Gaussian belief of the state vector is needed. In this example, we define the initial mean and covariance matrix of the state vector as follows. We assign a large variance for both the initial position and the speed so that the initial belief covers a wide range of possible values:

$$\mu_0 = \begin{bmatrix} 0 \\ 0 \end{bmatrix}, \quad \Sigma_0 = \begin{bmatrix} 1000^2 & 0 \\ 0 & 1000^2 \end{bmatrix}.$$

For demonstration purposes, we assume the measurement data are collected from a specific scenario called the "true scenario" described as follows. The agent starts from the initial position of 20.0 m and moves at a constant speed of 15.0 m/s all the time. The update of the agent's position in each step has noise that is sampled from $\gamma^p \sim N\left(0, 2.0^2\right)$, which is the same as that defined in the state transition model. The state values obtained from this true scenario is referred to as the "true state," which needs to be estimated. The measurement data are noisy and is generated using Equation (6.65) based on the true state values.

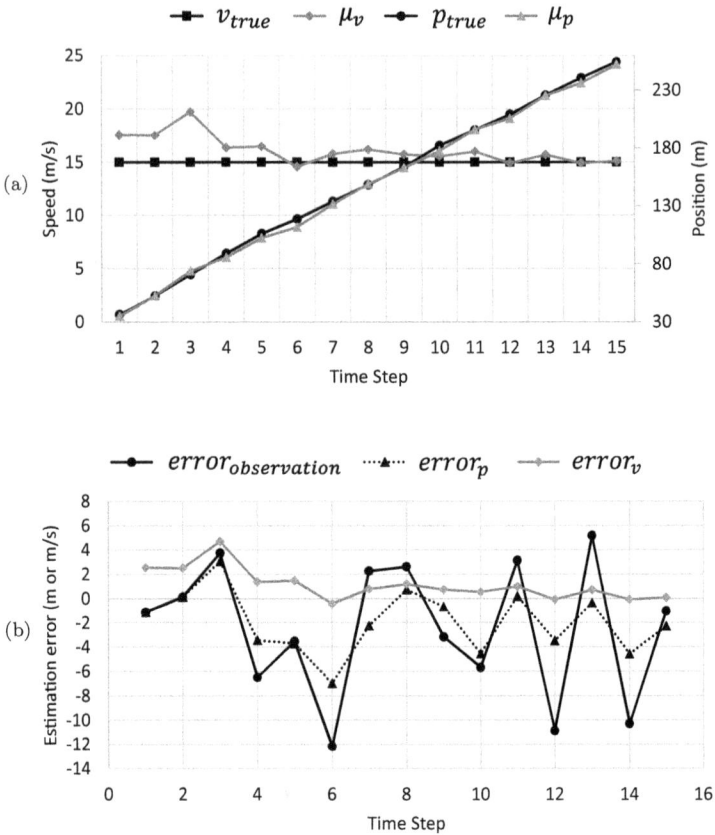

Figure 6.13.   State estimation results using the Kalman filter: (a) true state and estimated state; (b) observation error and estimation errors.

Using the Kalman filter algorithm described in Algorithm 6.2, the mean and covariance of the state vector (including position x^p and speed x^v) are computed in a stepwise fashion. Figure 6.13 shows the state estimation results for the first 15 steps of the Kalman filtering. For simplicity, the figure only shows the means of the estimated position and speed, denoted by μ_p and μ_v, respectively. Figure 6.13(a) shows the estimated μ_p and μ_v in comparison with the true position and speed (p_{true} and v_{true}). It can be seen that both μ_p and μ_v are able to converge to their true states over time. Figure 6.13(b) shows the difference between μ_p and p_{true} (denoted by $error_p$), the difference between μ_v and v_{true} (denoted by $error_v$), and the difference between the measurement data y and p_{true} (denoted by

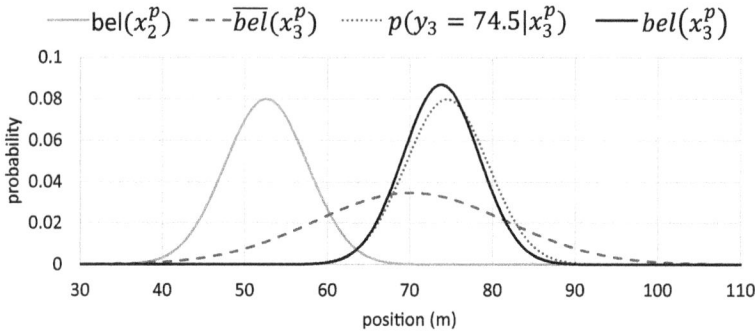

Figure 6.14. Prior distribution, likelihood, and posterior distribution in step 3 of the Kalman filtering.

$error_{observation}$). It can be seen that in almost all steps, $error_p$ is smaller than $error_{observation}$. In other words, the Kalman filter is able to estimate the true position more accurately than that indicated by the measurement data collected from the GPS sensor. This is because the position estimate from the Kalman filter incorporates not only information from the measurement data but also information from the state transition model.

Figure 6.14 illustrates how the Kalman filter estimates the belief for position x^p from the sequential Bayesian filtering point of view. The figure shows the step 3 of the Kalman filtering. In this step, the measurement data $y_3 = 74.5$. The posterior belief from step 2 (denoted by $bel(x_2^p)$) is shown by the gray Gaussian curve on the left. This belief evolves to the prior belief $\overline{bel}(x_3^p)$ after the agent moves to the right (which introduces uncertainty), which is shown as the dashed curve. The likelihood $p(y_3 = 74.5|x_3^p)$ of observing measurement data 74.5 is shown by the dotted-line curve. The Kalman filter algorithm integrates the information from $\overline{bel}(x_3^p)$ and $p(y_3 = 74.5|x_3^p)$ to produce the posterior belief $bel(x_3^p)$ that is shown by the black curve.

6.8 Particle Filters

Particle filters, also called *sequential Monte Carlo (SMC)* methods, are a set of algorithms that use the Monte Carlo techniques to realize sequential Bayesian filtering. Particle filters use samples to represent beliefs. They are nonparametric filters, meaning that they do not rely on a fixed functional form of belief distribution. The nonparametric nature of particle filters makes them applicable to a wide range of systems, including systems with nonlinear non-Gaussian behavior and systems with discrete or hybrid states.

To understand how the particle filters work, we first describe two fundamental concepts of particle filtering: *importance sampling* and *resampling*.

6.8.1 *Importance sampling*

Particle filters represent a belief distribution using a set of samples, each of which is called a *particle*. A particle is a concrete instantiation of the state vector. Let $\{x_k^{(1)}, x_k^{(2)}, \ldots, x_k^{(N)}\}$ be the set of particles representing the posterior distribution $p(x_k|y_{1:k}, u_{1:k})$ at time step k, where N is the size of the particle set. As discussed in the sample-based representation (Section 6.3.2), if N particles are sampled directly from $p(x_k|y_{1:k}, u_{1:k})$, then the distribution $p(x_k|y_{1:k}, u_{1:k})$ can be approximated by

$$p(x_k|y_{1:k}, u_{1:k}) \approx \frac{1}{N} \sum_{i=1}^{N} \delta(x - x_k^{(i)}), \qquad (6.66)$$

where $\delta()$ is the Dirac delta function. The larger the N, the more accurate the approximation.

Unfortunately, it is usually impossible to draw samples directly from the posterior distribution $p(x_k|y_{1:k}, u_{1:k})$, which is unknown and needs to be estimated. To address this problem, importance sampling is used.

In importance sampling, the distribution that cannot be directly sampled is called the *target distribution*. While it is difficult to sample from the target distribution, one may sample from another distribution, called the *proposal distribution* or *importance distribution*, and then approximate the target distribution using *weighted samples* from the proposal distribution. Let $p(\mathrm{x})$ be the density function of the target distribution, $q(x)$ be the density function of the proposal distribution, and $x^{(i)}$ be the samples from the proposal distribution $q(x)$. Then, the target distribution $p(x)$ can be approximated by

$$p(x) \approx \sum_{i=1}^{N} w^{(i)} \delta(x - x^{(i)}), \qquad (6.67)$$

where $w^{(i)}$ is the normalized importance weight for sample $x^{(i)}$ and is computed as follows:

$$\overline{w}^{(i)} = \frac{p(x^{(i)})}{q(x^{(i)})}, \qquad w^{(i)} = \frac{\overline{w}^{(i)}}{\sum_{j=1}^{N} \overline{w}^{(j)}}. \tag{6.68}$$

In Equation (6.68), $p(x^{(i)})$ is the probability density of the target distribution at position $x^{(i)}$ and $q(x^{(i)})$ is the probability density of the proposal distribution at position $x^{(i)}$. Here, $\overline{w}^{(i)}$ is called the *importance weight* of $x^{(i)}$ and is calculated as the quotient between $p(x^{(i)})$ and $q(x^{(i)})$. $w^{(i)}$ is the *normalized (importance) weight* that is computed based on the normalization of $\overline{w}^{(i)}$ (the normalization makes $\sum_i w^{(i)} = 1$). In the following description, we use $w^{(i)} \propto \frac{p(x^{(i)})}{q(x^{(i)})}$ to denote that the normalized weight $w^{(i)}$ is computed from $\frac{p(x^{(i)})}{q(x^{(i)})}$.

The importance sampling can be applied to the filtering problem to approximate the posterior distribution $p(x_k|y_{1:k}, u_{1:k})$. Let $q(x_k|y_{1:k}, u_{1:k})$ be a proposal distribution that is related to the posterior distribution and from which one can draw samples. It can be shown that when $q(x_k|y_{1:k}, u_{1:k})$ satisfies certain properties (see Arulampalam *et al.* (2002) for more details), the posterior distribution $p(x_k|y_{1:k}, u_{1:k})$ can be approximated by a set of weighted particles drawn from the proposal distribution, and the weights of the particles can be computed in a recursive way. Specifically, let $\{x_k^{(1)}, x_k^{(2)}, \ldots, x_k^{(N)}\}$ be the set of particles drawn from the proposal distribution. Then, the posterior distribution $p(x_k|y_{1:k}, u_{1:k})$ is approximated by

$$p(x_k|y_{1:k}, u_{1:k}) \approx \sum_{i=1}^{N} w_k^{(i)} \delta\left(x - x_k^{(i)}\right), \tag{6.69}$$

where $w_k^{(i)}$ is the normalized importance weight for particle $x_k^{(i)}$ and is computed in a recursive way as follows:

$$w_k^{(i)} \propto w_{k-1}^{(i)} \frac{p\left(y_k \mid x_k^{(i)}\right) p\left(x_k^{(i)} \mid x_{k-1}^{(i)}, u_k\right)}{q\left(x_k^{(i)} \mid x_{k-1}^{(i)}, y_k, u_k\right)}. \tag{6.70}$$

The derivation of the above recursive formula can be found in the literature (see, e.g., Arulampalam *et al.* (2002) and Thrun *et al.*

(2005)) and is omitted here. It can be shown that as $N \to \infty$, the approximation of Equation (6.69) approaches the true posterior density $p(x_k|y_{1:k}, u_{1:k})$.

In Equation (6.70), $w_k^{(i)}$ and $w_{k-1}^{(i)}$ are the ith particle's weights from steps k and $k - 1$, respectively. This equation is significant because it shows that $w_k^{(i)}$ can be computed in a recursive way. The recursive computation is useful for the filtering problem because it means that one only needs to store particles' most recent values and importance weights, and discard the past history and trajectories that lead to these values.

A decision needs to be made regarding what distribution to choose for the proposal distribution $q(x_k|y_{1:k}, u_{1:k})$. In theory, the optimal proposal distribution should be the one that minimizes the variance of the importance weights. In practice, a common choice is to choose the *prior distribution* (i.e., the state transition density defined in Equation (6.8)) as the proposal distribution. In other words,

$$q(x_k|y_{1:k}, u_{1:k}) = p\left(x_k \,|\, x_{k-1},\, u_k\right). \tag{6.71}$$

In this case, substituting (6.71) into (6.70) yields

$$w_k^{(i)} \propto w_{k-1}^{(i)} p(y_k|x_k^{(i)}). \tag{6.72}$$

Choosing the state transition density as the proposal distribution has several advantages from the computation point of view. First, it is straightforward to implement because drawing a sample from the proposal distribution is realized by running the state transition model to obtain a new state value. Second, it significantly simplifies the weight computation, as shown in Equation (6.72): The new weight $w_k^{(i)}$ of a particle is computed by multiplying its previous weight $w_{k-1}^{(i)}$ with the likelihood probability $p(y_k|x_k^{(i)})$. This is intuitive to understand because a higher likelihood probability leads to a higher weight. Hereinafter, we consider the state transition density as the *default choice* for the proposal distribution in particle-filter-based data assimilation.

The importance sampling described above gives rise to a basic particle filtering algorithm, named the *sequential importance sampling (SIS)* particle filter, for estimating a posterior distribution from measurement data. Let $\chi_{k-1} = \{\langle x_{k-1}^{(i)}, w_{k-1}^{(i)}\rangle\}_{i=1}^{N}$ be the set of weighted particles that represents the posterior distribution at step $k-1$, where $\langle x_{k-1}^{(i)}, w_{k-1}^{(i)} \rangle$ is the ith weighted particle with a value of $x_{k-1}^{(i)}$ and an importance weight of $w_{k-1}^{(i)}$. Using the SIS particle filter, the posterior distribution $p(x_k | y_{1:k}, u_{1:k})$ of step k is approximated by the set of weighted particles $\chi_k = \{\langle x_k^{(i)}, w_k^{(i)} \rangle\}_{i=1}^{N}$, where $x_k^{(i)}$ is computed by sampling from the state transition density $p\left(x_k \mid x_{k-1}, u_k\right)$ and $w_k^{(i)}$ is computed using Equation (6.72). The sampling from the state transition density $p\left(x_k \mid x_{k-1}, u_k\right)$ for a particle $x_k^{(i)}$ is implemented by running the state transition model (e.g., a simulation model) starting from the state $x_{k-1}^{(i)}$.

In summary, the SIS particle filter updates particles' state values and importance weights in a stepwise fashion. In each step, it evolves each particle's state into a new state using the state transition model and then computes the particle's importance weight by multiplying its previous weight with the likelihood probability $p(y_k | x_k^{(i)})$ for the new state.

6.8.2 *Resampling*

A problem of the SIS particle filter is the *degeneracy* phenomenon: As the time step k increases, the distribution of the importance weights $w_k^{(i)}$ becomes more and more skewed. As a result, after some time steps, only one particle becomes dominant and all other particles have negligible weights. The particles with negligible weights fall into regions of low posterior probability and thus have little contribution to the approximation of the posterior distribution. The single dominant particle belongs to a region of high posterior probability; however, it cannot adequately represent the overall posterior distribution by itself.

To address the degeneracy problem, the basic SIS particle filter is extended by adding an additional step called *resampling*. The goal of resampling is to eliminate the particles with small weights and concentrate on the particles with large weights. The resampling step generates a new set of particles that are offspring particles drawn from the original particle set. This new set of particles still approximates the posterior distribution that is represented by the original particle set. Different strategies may be used to determine when to

invoke resampling. For example, resampling may be invoked only
when a significant degeneracy is observed (Arulampalam *et al.*, 2002).
In this book, we consider the simplest strategy that invokes resampling in every step.

Formally, the resampling step is to sample N particles from the
original particle set to approximate the discrete representation of the
posterior distribution given by $p(x_k|y_{1:k}, u_{1:k}) \approx \sum_{i=1}^{N} w_k^{(i)} \delta \left(x - x_k^{(i)} \right)$.
This is equivalent to selecting a number of offsprings from the particles proportionally based on their weights. Particles with large
weights are more likely to be selected, often multiple times, whereas
particles with small weights are less likely to be selected; some of the
particles will not be selected at all and thus be eliminated. The end
result of the resampling step is a set of N equally weighted particles.
Together, they approximate the posterior distribution as follows:

$$p(x_k|y_{1:k}, u_{1:k}) \approx \frac{1}{N} \sum_{i=1}^{N} N_k^{(i)} \delta \left(x - x_k^{(i)} \right), \tag{6.73}$$

where all particles have the same weight of $1/N$. Here, $N_k^{(i)}$ is the
number of offsprings of particle $x_k^{(i)}$; it is an integer number such
that $\sum_{i=1}^{N} N_k^{(i)} = N$. If $N_k^{(i)} = 0$, then the particle $x_k^{(i)}$ is eliminated;
if $N_k^{(i)} \geq 1$, then the particle $x_k^{(i)}$ is replicated to produce one or
more offsprings (the multiple offspring particles are exactly the same
duplication of $x_k^{(i)}$).

The resampling in every step brings an important consequence to
the weight computation. After resampling, because all the particles
have the same weight of $1/N$, the weight information does not need
to be explicitly passed to the next cycle of the computation. Thus,
Equation (6.72) can be simplified as

$$w_k^{(i)} \propto p(y_k|x_k^{(i)}). \tag{6.74}$$

Figure 6.15 illustrates two algorithms for implementing the resampling step, which are named as the *multinomial resampling* and *systematic resampling* (Hol *et al.*, 2006). Both algorithms need to first
construct a *cumulative weight sum array* to record the cumulative

(a) Multinomial resampling

(b) Systematic resampling

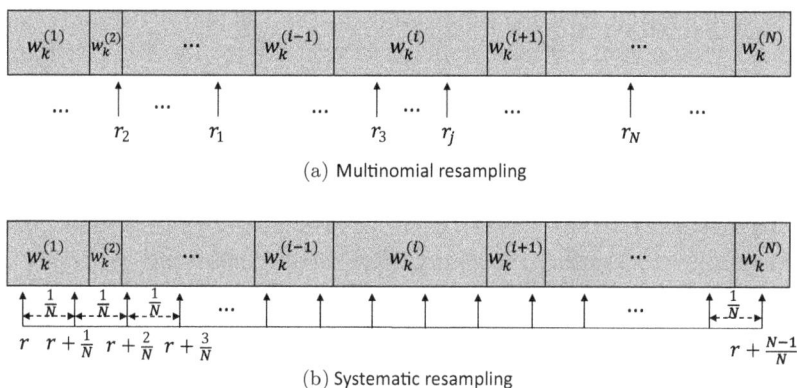

Figure 6.15.   Illustration of resampling algorithms.

sum of particles' weights, which is illustrated by the gray horizontal bar in the figures. Each particle occupies a segment of the bar; the length of the segment is defined by the particle's weight. Since particles' weights are normalized, the total length of the bar is 1.0. The multinomial resampling (illustrated by Figure 6.15(a)) generates N random numbers independently according to the uniform distribution between 0 and 1. It then selects the particles by checking which segments the random numbers fall in. For example, the random number r_2 falls in the segment of $w_k^{(2)}$, and thus, particle 2 would be selected. Similarly, random numbers r_3 and r_j fall in the segment of $w_k^{(i)}$, and thus, the ith particle would be selected twice based on these two random numbers. In general, since the random numbers are uniformly distributed, a longer segment would see more random numbers fall within it, and thus, the corresponding particle is more likely to be selected.

The multinomial resampling uses N uniformly distributed random numbers to select the particles. Since the goal of these N random numbers is to have N uniformly distributed data points to check against the cumulative weight sum array, a more systematic way is to generate those data points directly and make them evenly distributed. This leads to the systematic resampling illustrated by Figure 6.15(b). Instead of generating N random numbers, the systematic resampling generates a single random number in the interval of $[0, 1/N]$. It then computes the rest of the $N - 1$ data points by repeatedly adding the fixed amount of $1/N$ on top of the previous value. In this way, the overall N data points are evenly distributed across 0 and 1. The particles are then selected by checking which segments the N data points fall in. Since the N data points are evenly distributed and the segment lengths correspond to particles' weights, the particles would be selected proportionally based on their weights.

Algorithm 6.3 provides a concrete realization of the systematic resampling algorithm. The algorithm takes the input of a set of weighted particles, denoted by $\overline{\chi}_k = \{\langle \overline{x}_k^{(i)}, w_k^{(i)} \rangle\}_{i=1}^{N}$, where particle $\overline{x}_k^{(i)}$ is indexed by i and has a normalized weight of $w_k^{(i)}$. It returns a new set of particles, denoted by $\chi_k = \{x_k^{(j)}\}_{j=1}^{N}$. Note that after resampling, all the particles have the same weight of $1/N$. Thus, the new particle set does not need to explicitly list the weight value.

Algorithm 6.3: Systematic Resampling Algorithm.

Algorithm input: $\{\langle \overline{x}_k^{(i)}, w_k^{(i)} \rangle\}_{i=1}^{N}$

1. Construct the cumulative weight sum array:
   - Set $c_1 = w_k^{(1)}$
   - for $i = 2$ to N do
     $$c_i = c_{i-1} + w_k^{(i)}$$
     end for

2. Generate a random number between 0 and $1/N$, and then symmetrically increase it and check with the cumulative weight sum array to select particles:
   - Generate random number $r = r$ and $(0, 1/N)$
   - Set $\chi_k = \emptyset$
   - Set $i = 1$
   - for $j = 1$ to N do
     $$u = r + (j - 1)/N$$
     while $u > c_i$
        $i = i + 1$
     end while
     set $x_k^{(j)} = \overline{x}_k^{(i)}$ and add $x_k^{(j)}$ to χ_k
     end for

3. Return χ_k

Compared to the multinomial resampling, the systematic resampling algorithm has several advantages. First, it minimizes the Monte Carlo variation involved in the resampling procedure (because the N data points are evenly distributed). For example, if all particles have the same weight, the systematic resampling would select each particle exactly once. Second, it has a computation complexity of $O(N)$. This is due to the fact that both the cumulative weight sum array and the N data points are ordered as they are sequentially increased from the previous value. On the other hand, the multinomial resampling would require sorting the N random numbers or searching the cumulative weight sum array for each random number, resulting in a complexity of at least $O(N \log N)$. Due to the above reasons, the systematic resampling algorithm is preferred.

It is worth noting that while the resampling step addresses the degeneracy problem, it may cause a loss of diversity, as the resulting particle set may contain many duplicated particles. This is especially true if only one particle has a dominant weight — such a particle would be selected many times, making the new particle set dominated by a single particle. This is known as the *sample impoverishment* problem, which can be severe in the case of small process noise (Arulampalam *et al.*, 2002). Chapter 7 discusses a technique called *particle rejuvenation* that perturbs the resampled particles by adding noises to them.

6.8.3 *The particle filter algorithm*

Having introduced the importance sampling and the resampling concepts, we are ready to present the particle filter algorithm. Particle filters are a set of algorithms that vary in implementation details, such as the choice of proposal distribution and whether and when to carry out resampling. We focus on one of the most important particle filter algorithms called the *bootstrap filter* (Gordon *et al.*, 1993). The bootstrap filter implements a *sequential importance sampling with resampling (SISR)* procedure. It uses the prior distribution $p(x_k \mid x_{k-1}, u_k)$ (i.e., the state transition density) as the proposal distribution in importance sampling and invokes resampling in every step. Algorithm 6.4 shows the bootstrap filter, where k denotes the time step, N denotes the number of particles, and y_k^{md} denotes the measurement data at step k.

Note that Algorithm 6.4 does not strictly follows the prediction–update procedure, as in the sequential Bayesian filtering algorithm

(Algorithm 6.1). Instead, the algorithm is structured according to the importance sampling and resampling steps. This does not mean that the bootstrap filter does not follow the general prediction–update procedure. In fact, the first for loop of the importance sampling step is to generate N samples $\overline{x}_k^{(i)}, i = 1, \ldots, N$ using the state transition model. This corresponds to the prediction step in the prediction–update procedure. The rest of the computation, including the importance weight computation, normalization, and resampling, correspond to the update step of the prediction–update procedure.

Algorithm 6.4: Bootstrap Filter Algorithm.

1. Initialization, $k = 0$:

   - for $i = 1, \ldots, N$, sample $x_0^{(i)} \sim p(x_0)$.
   - Advance the time step by setting $k = 1$.

2. Importance sampling step (input: $\{x_{k-1}^{(i)}\}_{i=1}^N$, u_k, y_k^{md}):

   - for $i = 1, \ldots, N$, sample (i.e., predict) the new state $\overline{x}_k^{(i)}$ using the state transition model $p(x_k | x_{k-1}^{(i)}, u_k)$.
   - for $i = 1, \ldots, N$, compute the importance weight as the likelihood probability: $\overline{w}_k^{(i)} = p\left(y_k = y_k^{md} | \overline{x}_k^{(i)}\right)$. See Section 6.6.1.
   - for $i = 1, \ldots, N$, normalize the importance weights: $w_k^{(i)} = \dfrac{\overline{w}_k^{(i)}}{\sum_{j=1}^N \overline{w}_k^{(j)}}$.

3. Resampling step (input: $\{\langle \overline{x}_k^{(i)}, w_k^{(i)} \rangle\}_{i=1}^N$).

   - Resample N particles $\{x_k^{(i)}\}_{i=1}^N$ according to the normalized weights using the resampling algorithm (Algorithm 6.3).

4. Advance the time step and start a new cycle:

   - Set $k \to k + 1$, and go to step 2.

Figure 6.16 illustrates one step of the bootstrap filter algorithm from the prediction–update procedure point of view. In the figure, $bel(x_{k-1})$ is the posterior distribution from step $k - 1$ and is represented by the set of particles from step $k - 1$. Each particle in the

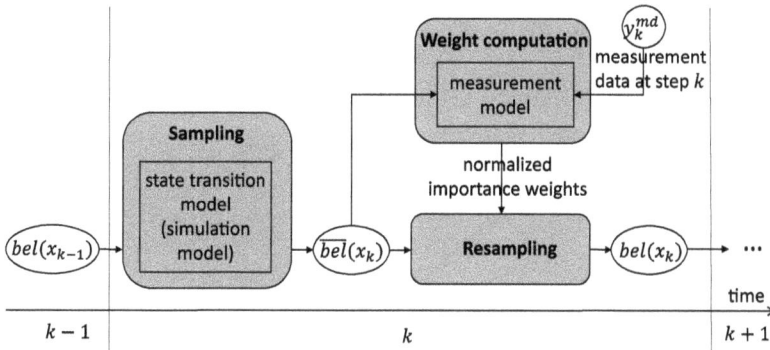

Figure 6.16. Illustration of the bootstrap filter algorithm.

set goes through the sampling step to predict its next state using the state transition model. The resulting particle set represents the prior distribution (denoted by $\overline{bel}(x_k)$) for step k. The importance weights of those particles are then computed using the measurement data and the measurement model, based on which the resampling step generates a new set of particles to represent the posterior belief $bel(x_k)$.

The initialization step of Algorithm 6.4 needs to generate N initial particles following the belief of the initial state. Similar to previous discussions, when knowledge about the initial state is available, the initial set of particles can be generated using that knowledge. Otherwise, a common practice is to generate the initial particles randomly covering a wide state space in a uniform way.

Figure 6.17 is a graphic illustration of how the particles and their associated weights are updated in the bootstrap filter algorithm. The horizontal axis at the bottom of the figure represents the state space x, and the small circles on top of it stand for the particles that have different state values from step $k-1$. We consider eight particles in this illustrative example (i.e., $N = 8$), which are denoted by $x_{k-1}^{(i)} (i = 1, \ldots, 8)$. These particles all have the same weight of $1/N$, illustrated by the same size of the circles. This set of particles (and their associated weights) provides an approximation of the posterior distribution from step $k-1$. This is indicated by $\{x_{k-1}^{(i)}, 1/N\}$ on the right side of the figure.

In the beginning of step k, these particles first go through the sampling step that predicts each particle's new state using the state

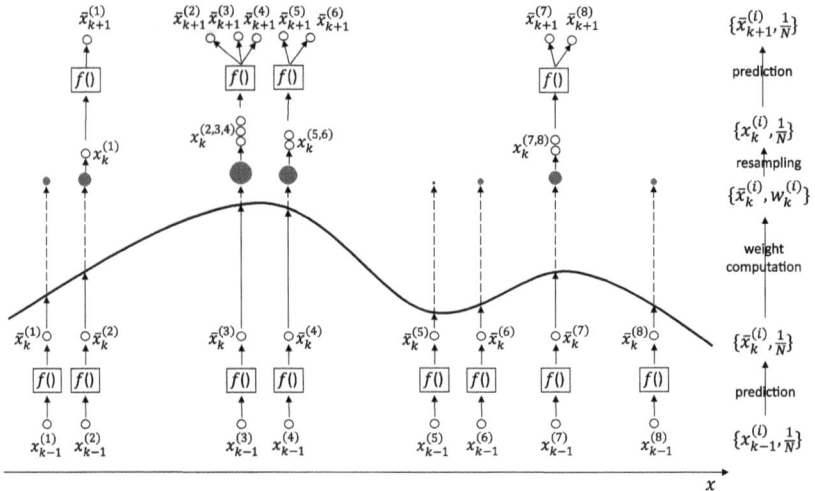

Figure 6.17. Illustration of how particles and their importance weights are updated in the bootstrap filter algorithm. This figure is adapted from Doucet *et al.* (2001).

transition model (the $f()$ box in the figure). The results of this prediction are denoted by $\overline{x}_k^{(i)}$ ($i = 1, \ldots, 8$) in the figure. Together they represent the prior distribution of the state. Then, the importance weights of these particles are computed according to their likelihood probabilities of observing the measurement data at this step (the likelihood probability curve for the state space is illustrated by the black curve in the figure). This results in a set of weighted particles $\{\overline{x}_k^{(i)}, w_k^{(i)}\}$, which acts as an intermediate approximation for the posterior distribution at step k. The different weights of the particles are shown by the different sizes of the filled circles corresponding to the particles. In this example, $\overline{x}_k^{(3)}$ has the largest weight, $\overline{x}_k^{(5)}$ has the smallest weight, and the other particles have weights that lie in between them.

Subsequently, the resampling step selects the particles based on their importance weights. The weights of the resulting particles (denoted by $x_k^{(i)}$ ($i = 1, \ldots, 8$)) are reset back to $1/N$, as illustrated by the same-sized small circles in the figure. This set of equally weighted particles is the final approximation of the posterior distribution at step k. Note that through resampling, the particles with large weights are selected multiple times, while the particles with small weights may be eliminated. In this example, $\overline{x}_k^{(3)}$ has the largest weight and is selected three times, producing three offspring particles. These three offspring particles are denoted by $x_k^{(2)}$, $x_k^{(3)}$, and $x_k^{(4)}$, which all have the same value of $\overline{x}_k^{(3)}$. Similarly, $\overline{x}_k^{(4)}$ produces two offsprings (denoted by $x_k^{(5)}$ and $x_k^{(6)}$); $\overline{x}_k^{(7)}$ produces two offsprings (denoted by $x_k^{(7)}$ and $x_k^{(8)}$); and $\overline{x}_k^{(2)}$ produces one offspring (denoted by $x_k^{(1)}$). The other particles are not selected and thus eliminated. As the result of the resampling step, the new set of particles at step k has one copy of $\overline{x}_k^{(2)}$, three copies of $\overline{x}_k^{(3)}$, two copies of $\overline{x}_k^{(4)}$, and two copies of $\overline{x}_k^{(7)}$.

Then, at the subsequent step $k+1$, the set of particles from step k goes through the sampling step to evolve to new state values based on the state transition model. The results are denoted by $\overline{x}_{k+1}^{(i)}$ ($i = 1, \ldots, 8$) in the figure. The sampling step involves uncertainty due to the process noise of the state transition model. This means the particles that start from the same state value will result in different state values after the state transition. Consider the particles $x_k^{(2)}$, $x_k^{(3)}$, and $x_k^{(4)}$ as an example. These particles have the same state value before the state transition but end up with different prediction results (denoted by $\overline{x}_{k+1}^{(2)}$, $\overline{x}_{k+1}^{(3)}$, and $\overline{x}_{k+1}^{(4)}$, respectively). Afterward, the predicted particles will compute their importance weights and go through the resampling in the same way as described earlier, which is not shown in the figure.

The bootstrap filter algorithm has several attractive properties from the data assimilation point of view. First and foremost, the fact that it chooses to use the state transition model as the proposal distribution in importance sampling matches well with the prediction–update procedure that uses the state transition model to predict new state. This choice also greatly simplifies the weight computation, as shown in Equation (6.72). Second, the algorithm has a modular structure. The use of the state transition model and the measurement model in the algorithm makes it possible to treat each of them as a black box. This means, when applying the algorithm to different applications, one need only "plug in" the corresponding models related to the applications while keeping other parts of the algorithm unchanged.

The performance of particle filters depends on many factors. On the application aspect, the complexity of the state space (e.g., dimension of the state vector), the quality of the state transition model and measurement model, as well as the scale of process noise and measurement noise all have an impact on the particle filtering results. On the algorithm aspect, an important parameter of the algorithm is the number of particles. In general, particle filters need to use a large number of particles in order to ensure that the state estimate converges to the true state. Nevertheless, a large number of particles would increase the computation cost. This is especially true for complex simulation applications because in every data assimilation step, each particle needs to invoke a simulation to predict a new state. Advanced techniques (e.g., parallel and distributed computing) would be needed when dealing with complex applications that require a large number of particles.

6.8.4 *Illustrative example*

To illustrate the particle-filter-based data assimilation, we consider a similar example as the one described in Section 6.7.4 but with a more complex measurement model. The example includes a mobile agent moving on a one-dimensional road with a predefined speed that is unknown. The precise position x^p and the speed x^v are unknown and need to be estimated. The state transition model of the agent has the same structure as the one defined by Equation (6.63) and is restated here:

$$x_k^p = x_{k-1}^p + \Delta t x_{k-1}^v + \gamma^p,$$
$$x_k^v = x_{k-1}^v + \gamma^v, \qquad (6.75)$$

where x_{k-1}^p, x_k^p, x_{k-1}^v, and x_k^v are the agent's position and velocity at time steps $k-1$, and k, Δt is the time step interval and is set to be 1 s, γ^p is the position noise, and γ^v is the speed noise. In this example, both γ^p and γ^v are zero-mean Gaussian noises with $\gamma^p \sim N\left(0, 1.5^2\right)$ and $\gamma^v \sim N\left(0, 0.5^2\right)$. One can see that the state transition model defined by Equation (6.75) is a linear Gaussian model.

Different from the example in Section 6.7.4 that uses a GPS sensor, in this example, the agent carries a temperature sensor. The temperature sensor measures the temperature at the agent's position

(a) One-dimensional road and fireplace positions

(b) Temperature measurement data

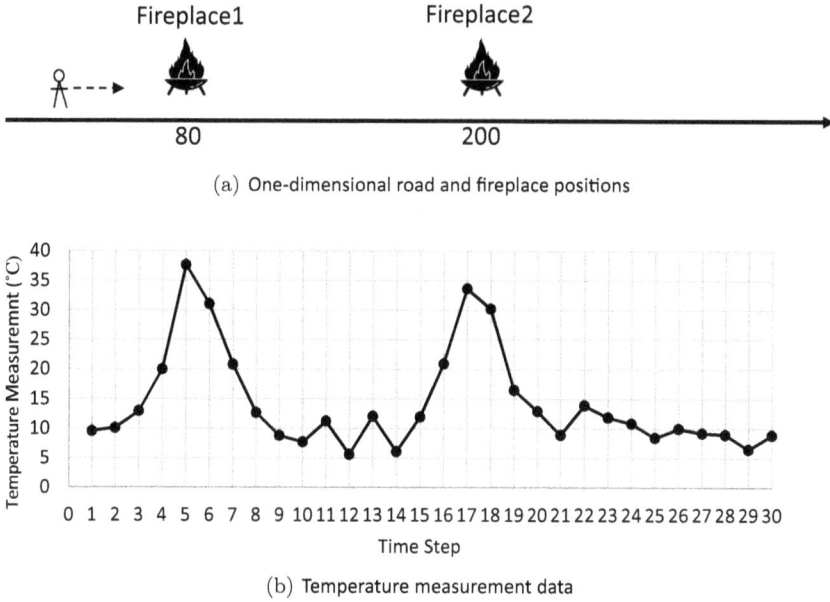

Figure 6.18.   The (a) mobile agent system and (b) measurement data.

every second. The measurement data are noisy due to the constraint of the sensor's precision as well as environmental noises. It is known that there are two outdoor fireplaces (called *Fireplace1* and *Fireplace2*) located at known positions on the road. Fireplace1 is located at the position of 80 m, and Fireplace2 is located at the position of 200 m, as shown in Figure 6.18(a). Due to the influence of the two fireplaces, the temperatures of different positions of the road are different.

The temperature at a position is influenced by how far the position is from the two fireplaces. Based on the temperature equation (Equation (6.32)) described in Section 6.6.1, we define the measurement model of this example as follows:

$$y_k = \max(y1_k, y2_k) + \varepsilon_k, \tag{6.76}$$

$$y1_k = T_c e^{\frac{-(x_k - 80.0)^2}{\sigma_c^2}} + T_a, \tag{6.77}$$

$$y2_k = T_c e^{\frac{-(x_k - 200.0)^2}{\sigma_c^2}} + T_a, \tag{6.78}$$

where $y1_k$ and $y2_k$ are the temperatures due to the influence of Fireplace1 and Fireplace2, respectively. The measurement y_k is the larger one between $y1_k$ and $y2_k$ (as defined by the $max()$ function) plus the measurement noise ε_k, which is a zero-mean Gaussian noise $\varepsilon_k \sim N(0, 2.0^2)$. In Equations (6.77) and (6.78), T_c is the temperature rise above the ambient temperature at the fireplace position and is set to be 30°C for both fireplaces, $T_a(°)$C is the ambient temperature and is set to be 10°C, and σ_c is a constant and is set to be 15 m.

The goal of data assimilation is to estimate the agent's dynamically changing state (i.e., position and moving speed) based on the temperature measurement data. In this example, the state vector of the agent is $x = [x^p, x^v]^T$, where x^p is the position and x^v is the speed. The state transition model is described by Equation (6.75), which is a linear Gaussian model. The measurement data are the noisy temperature data that are measured every second. These are scalar Gaussian measurement data. However, the measurement model, which is described by Equations (6.76)–(6.78), is a nonlinear model. Due to the nonlinearity of the measurement model, the Kalman filter is not applicable here. Instead, we use a particle filter to carry out the data assimilation.

For demonstration purposes, we assume the measurement data are collected from a scenario called the "true scenario" described as follows. The agent starts from an initial position of 20 m and moves at a constant speed of 10 m/s. The update of the agent's position in each step has a noise that is sampled from $\gamma^p \sim N\left(0, 1.5^2\right)$. The state values computed from this true scenario are called the "true state" and are unknown to the particle filter. The measurement data are noisy and are generated using Equations (6.76)–(6.78) based on the true state values. Figure 6.18(b) shows the measurement data of this example.

We use the bootstrap filter algorithm (Algorithm 6.4) with 2,000 particles to carry out the data assimilation. The particles are initialized by sampling their positions from a uniform distribution between 0 and 500 m, and their speeds from a uniform distribution between 0 and 50 m/s. Figure 6.19(a) shows the mean (denoted by μ_p) and standard deviation (denoted by σ_p) of the estimated position in comparison with the true position (denoted by p_{true}) for the first 30 steps. The mean and standard deviation are computed using information from all the 2,000 particles. Figure 6.19(b) compares

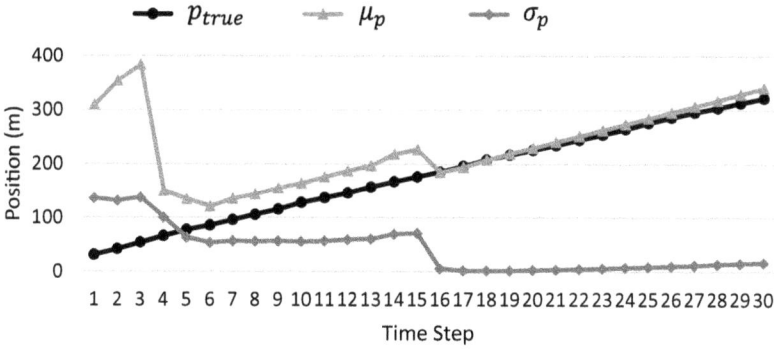

(a) True position and estimated position

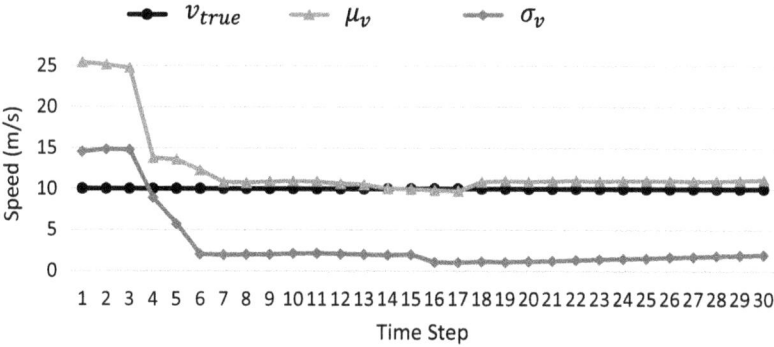

(b) True speed and estimated speed

Figure 6.19. State estimation results using a particle filter. (a) True position and estimated position and (b) true speed and estimated speed.

the mean (denoted by μ_v) and standard deviation (denoted by σ_v) of the estimated speed in comparison with the true speed (denoted by v_{true}).

As can be seen, in the first three steps, there are large errors between the estimated positions/speeds and the true positions/speeds. The standard deviations of the estimated positions/speeds in these three steps are also large, meaning that the estimates from different particles vary significantly. Starting from step 4, the difference between μ_p and p_{true} reduces significantly and then maintains at about the same level until step 15. The relatively large σ_p for the steps between step 4 and step 15 mean that the particles have not converged to p_{true} yet during these steps (the reason will be revealed in the following). After step 15, the difference between μ_p and p_{true} reduces to close to zero and then stays at a relatively small level. σ_p also drops to close to zero, indicating that the estimates from the different particles have converged. For the speed estimate, the difference between μ_v and v_{true} reduces significantly too, starting from step 4. It further reduces to close to zero at step 7 and maintains at a relatively small level. A similar pattern can be seen for σ_v: It keeps reducing and then stays at a relatively small level starting from step 6. This means, starting from step 6, the speed estimates from the different particles converge to the true speed.

To reveal more details about the estimates from the particles over time, Figure 6.20 shows the histogram of the 2,000 particles' position estimates at the end of step 1, step 4, step 15, and step 16 (all after resampling). For each step, the figure also marks the agent's true position (indicated by the cross sign) as well as the locations of the two fireplaces. To save space, we do not show the histogram for the speed estimates here.

The step 1 figure shows that at the end of step 1, the position estimates scatter all over the place except for the positions near the two fireplaces. The positions close to the two fireplaces are excluded from the estimation because the temperature measurement y_1 is $9.5°C$ (see Figure 6.18(b)). The likelihood of observing such a low temperature at the positions close to the fireplaces is very low. Thus, the particles whose positions are close to the two fireplaces have small importance weights and are eliminated by the resampling step.

At the end of step 4, the position estimates concentrate on the two regions that are on the left of Fireplace1 and Fireplace2. One of these regions corresponds to the agent's true position. Nevertheless, there is not enough information for the particle filter to distinguish between the two regions because they would both produce measurement data that match the measurement data received up to this time point. This is a case of bimodal distribution, which means the probability distribution has two modes (i.e., two peaks), which correspond to the two regions shown in the step 4 figure. In the following, we refer to the two modes as the *left mode* and the *right mode*.

Between step 4 and 15, the belief of the agent's position remains to be a bimodal distribution because the measurement data cannot distinguish between the two modes yet. Note that at step 15, the two modes shifted to the right compared to those at step 4. This is

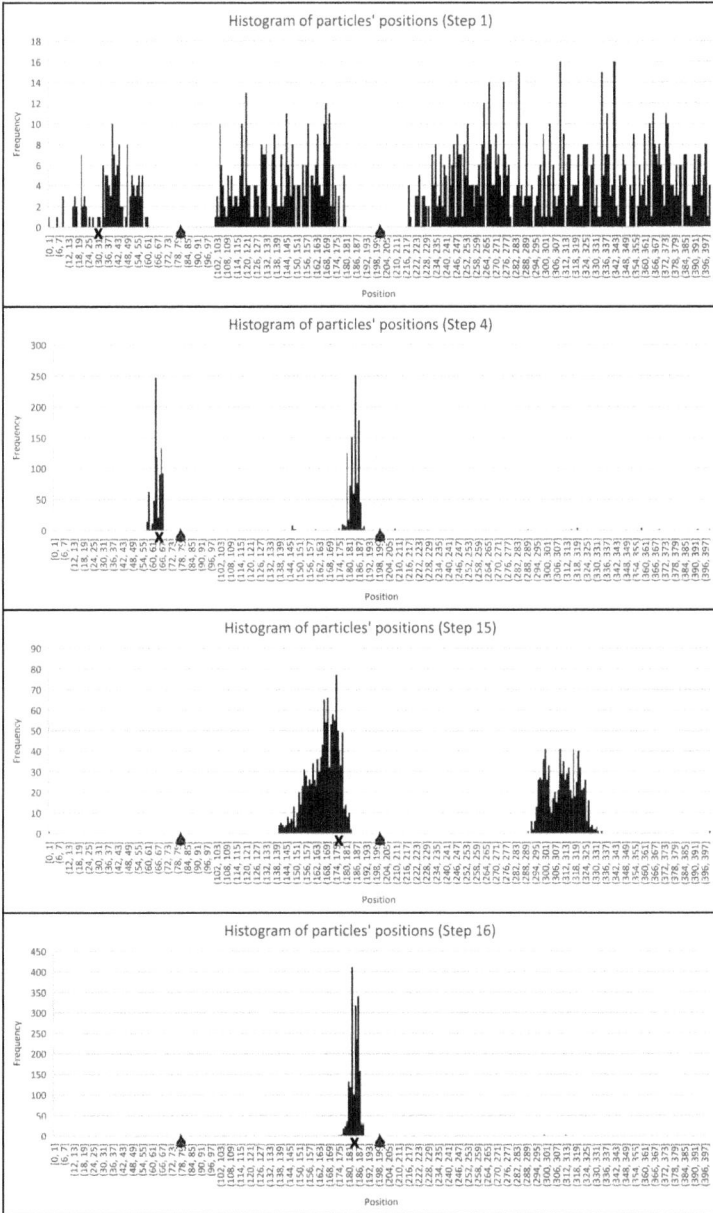

Figure 6.20. Histogram of position estimates over time: The true position is shown by the cross sign and the positions of the fireplaces are shown by the fireplace icons.

due to the position updates based on the estimated moving speeds. Furthermore, the two modes have larger variances compared to those at step 4, as shown by the larger width of each mode in the histogram. The larger variances are due to the process noises that are involved in the position updates over time.

At step 16, the temperature measurement y_{16} is 20.1°C. In this case, the right mode is eliminated because the likelihood probabilities for the positions in the right-hand region to observe a 20.1°C measurement are very low. As a result, the estimates from the particle filter concentrate only on the left-hand region, which is the region corresponding to the true position of the agent.

6.9   Sources

Several sections of this chapter are influenced by Thrun *et al.* (2005). These include the derivation of the prediction–update procedure in Section 6.5.1 and the description of the resampling algorithms in Section 6.8.2.

Chapter 7

Dynamic Data-Driven Simulation for Discrete Simulations

7.1 Introduction

Having described the dynamic data-driven simulation (DDDS) activities in Chapter 5 and the data assimilation methods in Chapter 6, this chapter focuses on the application of DDDS to *discrete simulations*. The term discrete simulation is used here to differentiate from *continuous simulations*, whose models are specified by ordinary or partial differential equations. Data assimilations for differentiate equation models have been studied in other science fields, such as meteorology, oceanography, and hydrology. The majority of those works are based on Kalman filter or its extensions, such as extended Kalman filter or ensemble Kalman filter. On the other hand, discrete simulations cover discrete event simulations and discrete time simulations (including many of the agent-based simulations). These simulations often deal with discrete or hybrid states (i.e., having both discrete and continuous state variables) and have nonlinear non-Gaussian behaviors, which makes the Kalman-filter-based data assimilation methods not applicable.

This chapter starts by presenting a framework of DDDS based on particle filters. It then discusses several practical issues and treatments for applying particle filters to data assimilation for discrete simulations. A significant portion of the chapter is devoted to a tutorial example based on a discrete event simulation model. The example demonstrates how the different DDDS activities work together to enable simulation-based prediction/analysis. It also offers a step-by-step tutorial on how to implement particle-filter-based data assimilation for a discrete event simulation model.

7.2   A Dynamic Data-Driven Simulation Framework

With advances in simulation technologies, there is a growing interest in using simulation to support real-time decision-making for dynamic systems. The operational use of simulation models in a real-time context requires simulation models to have the capability of providing real-time prediction/analysis for a system under study. Systematic ways of incorporating real-time data into simulation models are essential for achieving this goal. Figure 7.1 shows a framework of DDDS for supporting real-time decision-making. The framework includes multiple components that are organized in a layered architecture.

The ultimate goal of the framework is to support real-time decision-making for a system under study (the top layer). To achieve this goal, simulation-based prediction/analysis for the system's future behavior is needed (the prediction/analysis layer). The simulation-based prediction/analysis relies on an accurate assessment of the real-time condition of the system under study. This calls for dynamic state estimation so that the simulation can be

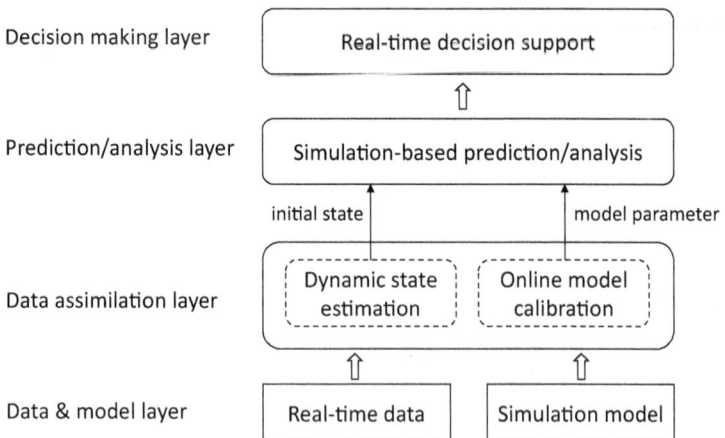

Figure 7.1.   A DDDS framework.

initialized from the estimated system state. Furthermore, simulation-based prediction/analysis needs the simulation model to accurately characterize the system in operation. This calls for online model calibration for the model parameters based on real-time data collected from the system. Both dynamic state estimation and online model calibration are addressed by data assimilation methods (the data assimilation layer). Data assimilation combines information from the real-time data and the simulation model. The former is the measurement data reflecting the system's real-time condition; the latter is the dynamic model describing how the system changes its state over time. The real-time data and simulation model are part of the model & data layer, which is the foundation of the overall framework.

This framework highlights the central role of data assimilation and shows the dependencies among the different components involved in DDDS. Note that the DDDS activity of *external input modeling & forecasting* is not shown in this framework. Modeling/forecasting a system's future inputs is important for simulation-based prediction/analysis if such inputs need to be modeled and forecasted (see Chapter 5 for more details). In this chapter, we assume that the inputs are either known or already modeled as part of the simulation model, and thus, we leave the *external input modeling & forecasting* activity out of the discussion. The components of the framework are described as follows:

- **Real-time data:** Real-time data are real-time measurement data collected from the dynamic system in operation. These measurement data are collected by sensors deployed within the dynamic system. Typically, real-time data are noisy data subject to the accuracy and precision constraints of the sensors.
- **Simulation model:** The simulation model is a dynamic model modeling how the dynamic system works. To support simulation-based prediction/analysis for a system in operation, a simulation model needs to be developed at the abstraction level corresponding to the actual operation of the system. Due to errors that commonly exist in simulation models, it is inevitable for a simulation model's result to deviate from a real system's behavior.
- **Data assimilation algorithm:** The data assimilation algorithm is the specific algorithm to carry out data assimilation. Due to the nonlinear non-Gaussian behavior and the discrete or hybrid states

of discrete simulations, we choose to use particle filters to carry out data assimilation for discrete simulations.

- **Dynamic state estimation:** Dynamic state estimation refers to the activity of estimating the dynamically changing state of a real system based on real-time measurement data collected from the system. The estimated state is then used to initialize simulation runs for simulation-based prediction/analysis. The dynamic state estimation is achieved through data assimilation.

- **Online model calibration:** Online model calibration refers to calibrating the parameters of a simulation model in an online fashion based on real-time measurement data collected from a dynamic system. The calibrated simulation model is then used to support more accurate simulation-based prediction/analysis. Online model calibration can be formulated as a joint state–parameter estimation problem and achieved through data assimilation.

- **Simulation-based prediction/analysis:** Simulation-based prediction/analysis refers to running simulations in real time to predict or analyze a dynamic system's future behavior. Simulation-based prediction/analysis utilizes the results of dynamic state estimation and online model calibration from the data assimilation layer.

- **Real-time decision support:** The real-time decision support is to use information from simulation-based prediction/analysis to support real-time decision-making for the dynamic system under study. How to carry out real-time decision support is application specific and is not the focus of this framework.

Several other things are worth mentioning for the above framework. First, simulation-based prediction/analysis may be invoked in each data assimilation step or in every few steps. Whenever it is invoked, it uses the data assimilation results from the most recent step to predict/analyze the future. Figure 7.2 shows how data assimilation and simulation-based prediction/analysis work together over time. In the figure, simulation-based prediction/ analysis is invoked in two consecutive data assimilation steps (step k and step $k + 1$). The data assimilation results of each step are used to set up the simulation-based prediction/analysis of that step. They also become the input for the next step's data assimilation.

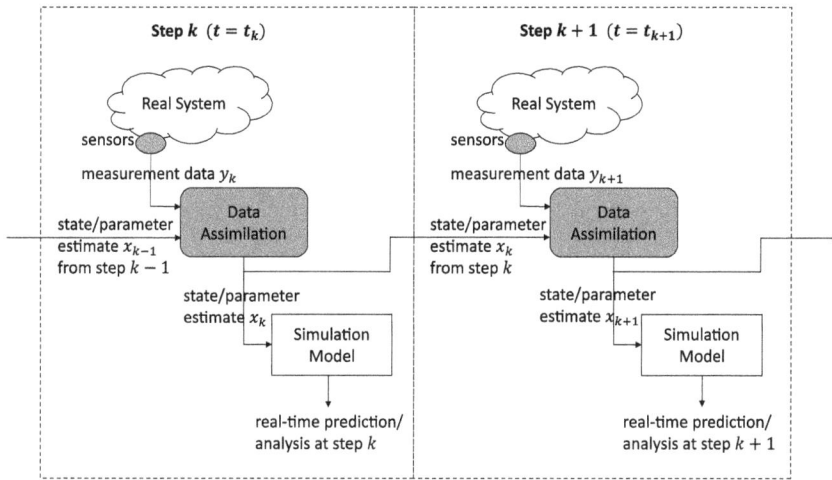

Figure 7.2. Data assimilation and simulation-based prediction/analysis over multiple steps.

Second, in particle-filter-based data assimilation, the state/ parameter estimations are represented by a set of particles. Each particle is a specific instance of the state/parameter estimate. When carrying out prediction/analysis, for each particle, a simulation would run using the initial state/model parameter defined by the particle. The prediction results from all the particles thus represent the distribution of the prediction results, based on which statistical moments, such as expectation and variance, can be computed. This is illustrated in Figure 7.3. In the figure, $x_k^{(i)}$ represents the state/parameter value of the ith particle at step k and $r_k^{(i)}$ is the prediction result based on the simulation run constructed from $x_k^{(i)}$. The simulation runs for the different particles have the same model structure. However, they start from different initial states and may use different parameter values (in the case of online model calibration).

Note that in Figure 7.3, each particle runs only one simulation and produces one prediction/analysis. If the simulation model is a stochastic model, different runs of the simulation starting from the same initial state will lead to different prediction results. In these cases, for each particle, multiple simulation runs may be invoked. The multiple simulation runs all start from the same initial state and

Data assimilation results at step k	Simulation-based prediction	Prediction results at step k

model parameter

$x_k^{(1)}$ ◯ —initial state→ [simulation model] —→ ◯ $r_k^{(1)}$

model parameter

$x_k^{(2)}$ ◯ —initial state→ [simulation model] —→ ◯ $r_k^{(2)}$

⋮   ⋮   ⋮

model parameter

$x_k^{(N)}$ ◯ —initial state→ [simulation model] —→ ◯ $r_k^{(N)}$

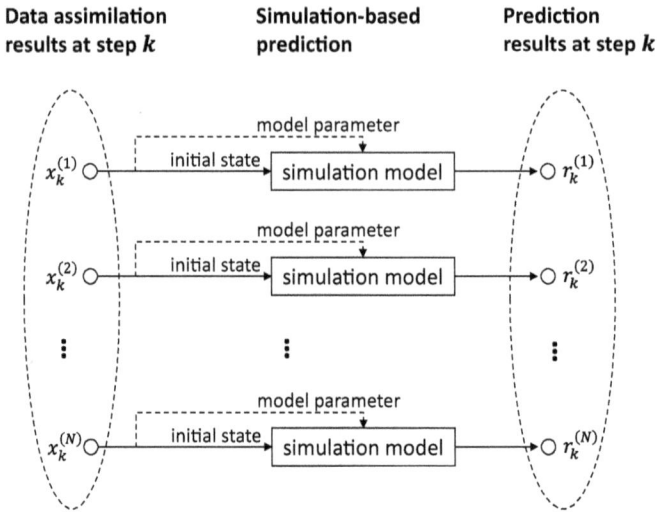

Figure 7.3.   Simulation-based prediction/analysis using particle filtering results.

use the same model parameters defined by the particle. The overall prediction/analysis results are then defined by all the simulation runs from all the particles.

Third, it is important to note that the simulation model servers two different roles in the framework. First, during data assimilation, it serves as the state transition model used by the particle filter algorithm to predict a prior distribution of the system state. Second, during simulation-based prediction/analysis, the simulation model is used to run simulations to predict/analyze a system's future behavior.

More specifically, during the importance sampling step of particle filtering, the purpose of the simulations is to evolve particles to the states of the current time. Whereas during simulation-based prediction/analysis, the purpose of the simulations is to start from the current time to predict the future. Due to their different purposes, the two simulations are configured differently. In the following description, we use $Sim\_DA()$ to refer to the simulation used in data assimilation and $Sim\_DDDS()$ to refer to the simulation used in simulation-based prediction/analysis. Considering the step $k+1$ in Figure 7.2 as an example, the differences between $Sim\_DA()$

and $Sim\_DDDS()$ are listed in the following. It is important to understand these differences in order to correctly configure the simulations used in data assimilation or simulation-based prediction/analysis:

- **Simulation start time and duration:** $Sim\_DA()$ starts from t_k and runs up to the current time t_{k+1}. $Sim\_DDDS()$ starts from t_{k+1} and runs into the future; how far into the future depends on how far one would like to predict/analyze the system's behavior.
- **Initial state:** $Sim\_DA()$ is initialized to the state estimate x_k, which is the data assimilation result from step k. $Sim\_DDDS()$ is initialized to the state estimate x_{k+1}, which is the data assimilation result from step $k+1$.
- **External inputs:** $Sim\_DA()$ uses the inputs (if any) that happened between t_k and t_{k+1}. These inputs already happened and are assumed to be known at time t_{k+1}. On the other hand, $Sim\_DDDS()$ uses inputs after t_{k+1}, which are future inputs that need to be modeled or forecasted.
- **Model parameters (in the case of online model calibration):** $Sim\_DA()$ uses the model parameters that are estimated from step k's data assimilation. $Sim\_DDDS()$ uses the model parameters that are estimated from step $k+1$'s data assimilation.

Finally, the length of a data assimilation step (i.e., $\Delta t_{k+1} = t_{k+1} - t_k$) is a design choice that varies for different applications. In the simplest scenario, a data assimilation is carried out whenever a new measurement becomes available. However, this may not be necessary when measurement data are collected in high frequency. A small data assimilation step can also bring about challenges in computation time because each step's data assimilation needs to be finished before the next step starts. On the other hand, a large data assimilation step means less frequent data assimilation. This can result in the measurement data not being utilized in a timely fashion. A larger data assimilation step also means it involves longer state transitions using the simulation model, which leads to larger uncertainties due to the process noise. One should consider all these factors when deciding how often to carry out data assimilation for a specific application.

7.3   Particle Filter-Based Data Assimilation

Several special treatments are needed when applying the bootstrap filter algorithm (Algorithm 6.4) to data assimilation for discrete simulations. This section describes these treatments.

7.3.1   *Sampling using discrete simulation models*

The bootstrap filter uses simulation models to generate samples of new state. To make discrete simulations and data assimilation work together in a structured way, one should pay attention to the following issues.

First, a clear identification of the state vector is essential for data assimilation. Nevertheless, many discrete simulations exist in computer programs and do not explicitly specify what state variables constitute the state. For example, a simulation program may use a large number of variables, only a subset of which actually define the system state. To carry out data assimilation for these models, a critical task is to explicitly identify the state vector and to define the mapping from the state vector to measurement data (i.e., the measurement model). In general, the state vector should include all the state variables (discrete and continuous), based on which a simulation can be constructed completely and precisely. For discrete event models that use time advances to schedule the next events, the elapsed times (or remaining times) should be part of the state vector unless they can be derived from other state variables. When a model has multiple components, the state vector needs to include the state variables from all the components.

Second, the sampling step involves running simulations to the current time and retrieving the states at the end of the simulations. Retrieving states is straightforward for a discrete time simulation because the updates of all the state variables are synchronized by the discrete time steps. Nevertheless, in a discrete event simulation, different model components update their states at different time instants according to the events occurring to them. The value of a state variable is not changed between the discrete event updates, and there is no inherent synchronization for updating all the state variables. On the other hand, data assimilation steps are time based. Thus, a mechanism is needed to synchronize the state of a discrete

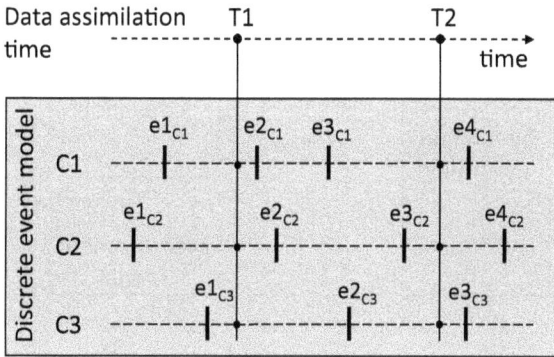

Figure 7.4. Updating and retrieving a discrete event model's state at data assimilation time steps.

event simulation based on the data assimilation time steps; otherwise, the state retrieved at a data assimilation step will be inaccurate.

Figure 7.4 illustrates this problem. In the figure, the gray box represents the discrete event model that has three component models: *C1*, *C2*, and *C3*. Each component goes through multiple events, which are illustrated by the black bars on the horizontal line (time line) associated with the component. In an uninterrupted simulation run, the simulation would proceed in the order of the timestamps of all the events, e.g., $e1_{c2}$, $e1_{c1}$, $e1_{c3}$, $e2_{c1}$, ... in this example. The simulation time jumps from one timestamp to the next timestamp. The state variables are updated only at those timestamp points when there are events, and the update is only for the involved components. On the other hand, a data assimilation step may happen at a time, such as $T1$ and $T2$ in Figure 7.4, that is not synchronized with any of the event time. This means we need to be able to stop the simulation at the data assimilation time ($T1$ and $T2$) and update all components' state variables to that time so that an accurate snapshot of the overall state can be retrieved.

How to update the state variables to a specific time instant is application specific. In general, for a continuous state variable, its value needs to be appropriately interpolated based on the time elapsed from the last update; for a discrete state variable, besides the most recent state value, the elapsed time since the last update also needs to be recorded as part of the state. The goal is to be able

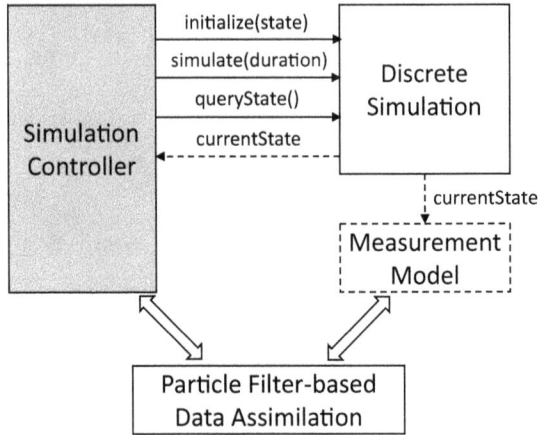

Figure 7.5. Simulation controller for data simulation.

to reconstruct a simulation precisely based on a recorded state value so that the next data assimilation step can start from the recorded states from the previous step.

Third, to make discrete simulations and data assimilation work together, it is useful to have a *simulation controller* to handle the interactions between data assimilation and simulation. The simulation controller acts as an intermediate component to bridge the data assimilation algorithm and the discrete simulation. Figure 7.5 shows this relationship. The data assimilation algorithm uses the simulation controller to initialize and run simulations and to retrieve state values. There are four types of messages/data between the simulation controller and the discrete simulation, as described in the following:

- The *initialize(state)* message is to initialize a simulation using a specific *state* value. This is needed at the beginning of each data assimilation step to set up simulations using the states represented by particles. The *state* parameter should include all the necessary information for initializing a simulation completely and precisely.
- The *simulate(duration)* message is to run the simulation for a specified duration, which is the time interval between the last data assimilation step and the current step. The simulation stops when the duration is reached.
- The *queryState()* message is to query the full state of the simulation model. The simulation controller sends this request at the end

of a simulation-based sampling. Upon receiving this request, the discrete simulation should update its state to the current time and return the full state using the *currentState* data described below.
- The *currentState* data is from the discrete simulation to the simulation controller to return the updated state to the simulation controller. The *currentState* is also used by the measurement model to compute predicted measurement.

The significance of the simulation controller is mostly at the conceptual level. Actual implementations of the simulation controller may vary significantly, ranging from developing a new module of simulation controller that is separated from the simulation program and data assimilation algorithm, to embedding the functions of the simulation controller in the existing code of simulation or data assimilation. Note that a discrete simulation may not be implemented in a way that readily supports all the interactions with the simulation controller. This means some extra work may be needed to restructure or extend the simulation code to support the interactions. The tutorial example described in Section 7.5.2 provides an example of how to extend a discrete event simulation model to support reporting the state at the time of data assimilation steps.

7.3.2 *Measurement data and measurement model*

Measurement data and measurement model are used to compute the likelihood probability that defines a particle's importance weight. A general approach to likelihood computation for scalar Gaussian measurement is described in Section 6.6. This section discusses several practical issues that often rise in data assimilation for discrete simulations.

Measurement data are collected by sensors. The sampling rate of sensors defines how often measurement data are collected. Our discussion so far has implicitly assumed that the rate of data assimilation (i.e., how often data assimilation is carried out) is the same as the rate of data collection. In other words, a data assimilation is carried out each time when a new measurement becomes available. While this makes sense in many cases, it may not be necessary or practical when sensors have a very high sampling rate. For example, a temperature sensor with a 4 Hz sampling rate would produce four temperature data per second. For systems whose temperature evolves slowly, a new data point would carry minimal new information. In this case, carrying out data assimilation for every new temperature measurement is not necessary. It may not be feasible either as the data assimilation may not be able to finish its computation before the next data point arrives. Due to this reason, it is common for data assimilation to be carried out at a slower rate than the rate of data collection.

A slower rate than the rate of data collection means that a data assimilation step may receive multiple measurement data points. The multiple data points need to be processed to provide a single measurement data used by the data assimilation step. Let t_{k-1}, t_k, and t_{k+1} be the data assimilation times for steps $k-1$, k, and $k+1$, respectively; $y_{k(1)}$, $y_{k(2)}$, ..., $y_{k(m)}$ be the m number of data points received between t_{k-1} and t_k; and $y_{k+1(1)}$, $y_{k+1(2)}$, ..., $y_{k+1(n)}$ be the n number of data points received between t_k and t_{k+1}. The m and n may have different values. Figure 7.6 shows four examples of how the multiple data points may be processed to produce the measurement data y_k and y_{k+1}. We note that besides these four examples, there exist many other ways of handling the data.

Figure 7.6(a) shows the case where there exists only one measurement data point and the data assimilation is carried out right at the time when the measurement arrives. In this ideal case, the one measurement data point is used directly. Figure 7.6(b) shows the case where the most recent data point is used as the measurement data (the other data points are discarded). An example of this is the GPS measurement data for a mobile agent — the newest GPS measurement is considered as the measurement data. Figure 7.6(c) shows the case where the measurement data is computed as the average of the multiple data points. An example of this is the traffic speed data for a road segment — the average speed computed from all the data points in the past time interval is used as the measurement data. Figure 7.6(d) shows the case where the measurement data is computed as the cumulative sum of all the data points. An example of this is the number of processed jobs from a manufacturing system — each data point describes a new batch of jobs, and the measurement data is the total number of jobs in the past time interval that is equal to the cumulative sum of all the data points.

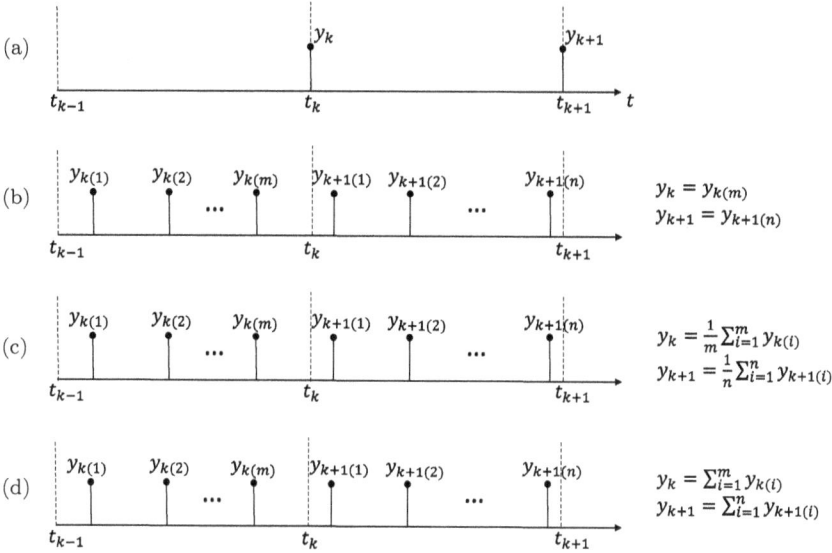

Figure 7.6. Examples of processing measurement data: (a) using the one measurement data point at the data assimilation time — this is the ideal case; (b) using the most recent data point; (c) computing the average of all the data points in the past time interval; (d) computing the cumulative sum from all the data points in the past time interval.

The above examples reveal several common errors that exist in the measurement data for data assimilation. First, there may exist a discrepancy between the data assimilation time and the measurement data collection time. This time discrepancy means that the measurement data used in data assimilation may not be exactly from the "current" time. This can be seen in the example of Figure 7.6(b), where the data assimilation at t_k uses the measurement data $y_{k(m)}$, which is collected before time t_k. Second, the processing of measurement data introduces errors too. For example, averaging the data points to produce measurement data introduces the possibility that the averaged result does not adequately reflect the most recent state changes. Besides the above errors, a sensor itself also has accuracy- and precision-related errors. In data assimilation, all these errors are modeled by the measurement noise.

A common choice for measurement noise is the zero-mean Gaussian noise. In the bootstrap filter, a particle's weight is defined by the likelihood probability, which is computed based on Equation (6.30) for scalar Gaussian measurement (see Section 6.6 for more details). The variance (denoted by σ^2) of the zero-mean Gaussian noise is thus an important factor because it directly influences the sensitivity of particles' important weights. While in principle, the σ^2 should be set based on the uncertainty involved in the measurement data, in practice, the precise uncertainty is unknown. Thus, it is common to tune σ^2 in an experiment in order to achieve the best data assimilation results. Generally, a small σ^2 would make the weight computation sensitive to the differences existing between particles. This makes it easy to differentiate between the good and bad particles. However, it also increases the possibility for only a few particles to have dominant importance weights and thus impairs the diversity of the particle population. On the contrary, a large σ^2 makes it easy to preserve the diversity of particles but can fail to differentiate the good particles from the bad ones. One should balance the sensitivity and diversity needs when tuning σ^2 in particles' weight computation.

A measurement model is a separate model from the simulation model. It models how the sensors produce measurement data based on the state of the system. For example, if ground temperature sensors are used to collect temperature data of a wildfire, the measurement model should model the deployment locations of the sensors and how each sensor's temperature is influenced by the fire. The measurement model takes the state produced by the simulation model as input and computes the measurement data. In some cases, the measurement data is a measurement of the events/activities (e.g., the number of processed jobs) occurred in a past time interval. For these cases, the measurement model can be implemented as an "observer" of the simulation to record the events/activities during a simulation run. An example of this can be found in the tutorial example in Section 7.5.2, which uses a *datasavingTransducer* model to collect the measurement data.

7.3.3　*Particle rejuvenation*

In particles filtering, a particle that has a large weight can generate many offsprings through the resampling step. If a discrete simulation model is deterministic or has small process noise, during the sampling

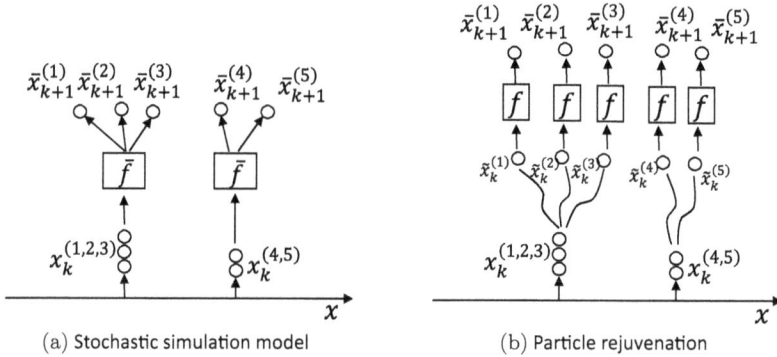

Figure 7.7. Improving the diversity of particles: (a) adding or increasing process noise to make the simulation model a stochastic model; (b) perturbing particles directly using particle rejuvenation.

step, all these offsprings will lead to the same or very similar states. This causes low diversity in the particle ensemble. One approach to address this issue is to make the simulation model a stochastic simulation model by adding or increasing the process noise associated with the state transition. This allows the particles that start from the same state to evolve to different states after a state transition. This is illustrated in Figure 7.7(a). In the figure, $x_k^{(1,2,3)}$ are three offspring particles resampled from the same particle, and their state transition results are denoted by $\bar{x}_{k+1}^{(1)}$, $\bar{x}_{k+1}^{(2)}$, and $\bar{x}_{k+1}^{(3)}$, respectively. A similar relationship applies to $x_k^{(4,5)}$ and their corresponding state transition results. The $\bar{f}$ box represents the modified simulation model after adding the process noise. The modified simulation model is used by the sampling step for state transition.

This approach has the advantage that the introduced process noise becomes an inherent part of the simulation model. However, it requires modifying the simulation program, which may not be straightforward because one needs to identify the places to introduce the process noise and to decide what process noise to add. Furthermore, a change in the simulation model raises concerns from the model validation point of view.

A more practical approach to address this issue is to employ a technique called *particle rejuvenation*. Particle rejuvenation directly adds stochastic perturbations (e.g., Gaussian noises) to the states represented by particles before the sampling step. This would avoid the situation of having identical particles to start the state transition. For discrete simulations that are deterministic or have small process noise, particle rejuvenation can improve the diversity of particles and make the data assimilation more robust. Figure 7.7(b) illustrates particle rejuvenation as an added step before simulation-based sampling. In the figure, the rejuvenated particles for $x_k^{(1,2,3)}$ are denoted by $\tilde{x}_k^{(1)}$, $\tilde{x}_k^{(2)}$, and $\tilde{x}_k^{(3)}$, respectively. Each of them has a different state value. After simulation-based sampling, their new states are denoted by $\bar{x}_{k+1}^{(1)}$, $\bar{x}_{k+1}^{(2)}$, and $\bar{x}_{k+1}^{(3)}$, respectively. Note that the simulation-based sampling is based on the unmodified simulation model (represented by the f box in the figure).

A different way of looking at particle rejuvenation is to treat it as part of an *extended state transition model* that first adds stochastic perturbations to the state variables and then uses the state transition model (i.e., the simulation model) to evolve the perturbated state.

A design question for particle rejuvenation is how much perturbation to add. On the one hand, the perturbation should not be too small so that the goal of improving the diversity of particles can be achieved. On the other hand, the perturbation should not overshadow the state change from the simulation model; otherwise, the evolution of particles' states is dominated by the noises added through perturbation.

Besides addressing the diversity issue described above, particle rejuvenation may also be used as a heuristic tool to compensate for the errors that exist between a model and a real system. In this role, the perturbation increases the possibility for the particles to cover the true state so that more robust data assimilation results can be achieved.

An example of particle rejuvenation is provided in the tutorial example in Section 7.5.3.

7.4   Identical Twin Experiment

The complexity of data assimilation calls for effective ways to evaluate a data assimilation project before applying it to a real system. This brings about the need for experiment design for data assimilation, which deals with how to set up experiments for data assimilation and evaluate its performance. A desirable way of experimenting with data assimilation is to run it against real-world scenarios, where measurement data collected from real systems are assimilated and the estimated states are compared with the real system states. Nevertheless, it is often impractical to set up such experiments. This is because the real measurement data (i.e., sensor data collected from real systems) are difficult to obtain or may not be available. Furthermore, the true system states are often unknown, and thus, it is impossible to compare the estimated states with the true states to evaluate.

To address the above issue, a popular approach is to employ the *identical twin experiment* design to set up experiments for data assimilation. In the identical twin experiment, a simulation (referred to as the "true system") is first run and the corresponding data are recorded. The measurement data obtained from this simulation are regarded as coming from the "true system," and the state trajectory recorded from this simulation is considered the "true" state (the quotation marks may be omitted later when the context is clear). Consequently, data assimilation is carried out using the measurement data, and then, the state estimates are checked against the true state. The first simulation that serves as the "true system" is often based on the same simulation model but uses different model configurations, such as different initial states, input trajectories, or random number seeds for generating behaviors, that are "unknown" to the data assimilation. In this way, it acts as a surrogate for the real system and generates synthetic measurements to be assimilated.

The identical twin experiment includes a "true system" simulation that is based on a specific configuration of the simulation model. To differentiate the model used in the "true system" simulation from the one used in data assimilation, we name the former as the *"true system" model* and the latter simply as the simulation model. Table 7.1 summarizes the differences between the two models.

The identical twin experiment has the advantage that the "true system" is known and the quality of the simulated measurements is under the complete control of the developer so that the performance and deficiency of data assimilation can be identified with much fewer

Table 7.1.   The "true system" model and the simulation model used in the identical twin experiment. Both models are based on the same simulation model but use different configurations.

Model name	The "true system" model	The simulation model
Role	Act as the real system.	Act as the imperfect state transition model used by data assimilation.
Simulation configuration	Use a specific configuration that is "unknown" to the data assimilation.	Use a configuration that is different from the "true system" model configuration.
Usage in data assimilation	(1) Generate "real" measurement data to be assimilated; (2) generate the "true" states that are needed for evaluating the data assimilation results.	(1) Serve as the state transition model that is needed in sequential Bayesian filtering; (2) in online model calibration, this is the model whose parameters need to be calibrated.

unknowns. It also allows for the testing of more scenarios than those for which real data are available.

Algorithm 7.1 describes the steps involved in carrying out particle-filter-based data assimilation using the identical twin experiment, where K is the total number of data assimilation steps and N is the total number of particles. This algorithm uses the same bootstrap filtering procedure described in Algorithm 6.4 but adds an extra step (Step 1) for setting up and simulating the "true system" model. Note that in the identical twin experiment, the number of data assimilation steps and the length of each step need to be defined before simulating the "true system" model so that the corresponding measurement data and true states can be recorded. Also, note that when carrying out data assimilation using measurement data from a real system, the only part that needs to be changed is to replace step 1 with the data collected from the real system.

Algorithm 7.1. Particle- Filter-based Data Assimilation using the Identical Twin Experiment.

1. Set up and simulate the "true system" model:
   - Decide the configuration for the "true system" model.
   - Set up and run a simulation using the "true system" model.
   - Record the measurement data y_k^{md} at the corresponding time $t = t_1, \ldots t_K$ for each data assimilation step k.
   - Record the "true" state x_k^{true} for each data assimilation step k (for evaluation purpose).

2. Data assimilation initialization, $k = 0$:
   - for $i = 1, \ldots, N$, sample $x_0^{(i)} \sim p(x_0)$.
   - Advance the time step by setting $k = 1$.

3. Importance sampling step:
   - for $i = 1, \ldots, N$, run a simulation using the simulation model starting from the initial state $x_{k-1}^{(i)}$ for a duration of $(t_k - t_{k-1})$, and record the resulting state $\bar{x}_k^{(i)}$. The simulation model has a configuration that is different from the "true system" model.
   - for $i = 1, \ldots, N$, compute the importance weight as the likelihood probability: $\bar{w}_k^{(i)} = p(y_k = y_k^{md}|\bar{x}_k^{(i)})$.
   - for $i = 1, \ldots, N$, normalize the importance weights.

4. Resampling step:
   - Resample N particles $\{x_k^{(i)}\}_{i=1}^N$ according to the normalized weights using the resampling algorithm (Algorithm 6.3).
   - Record the results for this step and compare the results with x_k^{true} for evaluation.

5. Advance the time step and start a new cycle:
   - If k reaches K, terminate; otherwise, set $k \leftarrow k + 1$ and go to Step 3.

7.5   A Tutorial Example

This section provides a tutorial example of DDDS based on a discrete event simulation model. The tutorial example serves multiple purposes, including: (1) demonstration of how different DDDS activities (e.g., dynamic state estimation and online model calibration) can work together to support simulation-based prediction/analysis for a "real world" application; (2) providing a step-by-step tutorial on how to implement particle-filter-based data assimilation for a discrete event simulation model; and (3) evaluation of the impact of several key parameters of particle filtering on data assimilation results. This tutorial example focuses on the state/parameter estimation for a single-component model. It complements the traffic data assimilation example (Hu and Wu, 2019) and the wildfire spread simulation application (presented in the following chapter), which focus on system-level state estimation involving many component models.

7.5.1   *The one-way traffic control system*

We consider a one-way traffic control system that is often used during road construction. The one-way traffic control system is illustrated in Figure 7.8. During a road construction, the one-way traffic control is managed by two persons deployed at the west and east ends of the one-way traffic road segment. Each person carries a STOP/SLOW handheld traffic paddle to control the traffic, where the STOP sign means vehicles should stop and wait and the SLOW sign means vehicles can slowly move ahead to pass the road segment. It is assumed that the two persons coordinate and always use the STOP/SLOW signs in opposite directions: When one uses the STOP sign, the other

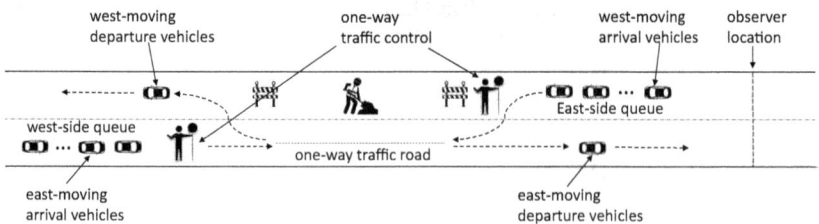

Figure 7.8.   The one-way traffic control system.

would use the SLOW sign. In the following description, we refer to the STOP sign as the red traffic light and the SLOW sign as the green traffic light, and we refer to the vehicles' moving directions as *east-moving* (moving toward east) and *west-moving* (moving toward west). During the time when the traffic light is green in a specific direction (east-moving or west-moving), the arrival vehicles moving in the opposite direction are queued. The queues at the west side and east side of the road segment are named as the *west-side queue* and *east-side queue*, respectively.

To ensure construction workers' safety, only one vehicle is allowed to move on the one-way traffic road at any time. During a green light period, the traffic-control person on the corresponding side would signal a vehicle to move ahead only when the previous vehicle has finished crossing the road segment. The one-way traffic control system uses the following rules to switch traffic lights:

1. If the elapsed time for the current moving direction (east-moving or west-moving) has reached a predefined threshold (120 s in this example) and the opposite moving direction has cars waiting, switch the traffic light. Note that the traffic light switches only after the road segment is cleared if there is a car already moving on the road.
2. If the current moving direction has no cars waiting and the opposite moving direction has cars waiting in the queue, switch the traffic light even if the 120 s threshold is not reached.
3. If the current moving direction has cars waiting in the queue and the opposite moving direction has no cars waiting, keep the traffic light unswitched even after the 120 s threshold. In this case, the cars on the current moving direction keeps moving forward.
4. If none of the current moving direction and the opposite moving direction has any cars needing to cross the road segment, keep the traffic light unswitched.

The vehicles at both sides of the road segment arrive randomly and independently, modeled by Poisson distributions. For the east-moving vehicles, the Poisson distribution has an arriving rate of $\lambda = 1/7$ (1 car per 7 s on average). For the west-moving vehicles, the arriving rate is $\lambda = 1/10$ (1 car per 10 s on average). This means there are more east-moving vehicles than west-moving vehicles. The time it takes for a vehicle to cross the road segment is also a random number, modeled by a truncated normal distribution that has a mean of $\mu = 4.0$ (seconds) and a variance of $\sigma^2 = 0.5^2$ and lies within the range of [3.0, 5.0].

To collect measurement data from the system, an observer (a sensor) is deployed at a location on the east end of the one-way traffic road that is marked as the "observer location" in Figure 7.8. The observer is able to count the number of moving vehicles crossing its location for both the east-moving departure vehicles (named as *eastMovDeparture*) and west-moving arrival vehicles (named as *westMovArrival*). The observer reports data every 30 s. It does not record the specific time that a vehicle crosses the observation location — all it reports is the total number of vehicles that have departed and arrived in the past time interval. The data reported by the observer is noisy (details to be described later).

7.5.2 *The discrete event simulation model*

The one-way traffic control system is modeled by a discrete event simulation model based on the DEVSJAVA environment (Zeigler and Sarjoughian, 2003). Figure 7.9 depicts the simulation model, which is a DEVS coupled model, including three atomic models (shown as gray boxes). The fourth atomic model *dataSavingTransducer* (the white box) is used to save data for data assimilation.

The three DEVS atomic models for the one-way traffic control system are described as follows (the specific code of the models are omitted here):

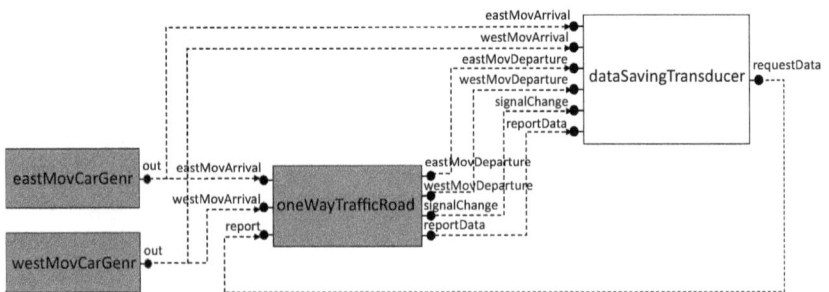

Figure 7.9.   Simulation model for the one-way traffic control system.

- *eastMovCarGenr*: generates the east-moving traffic arriving at the west side of the road segment. Traffic is generated based on a Poisson distribution with a rate of $\lambda = 1/7$.
- *westMovCarGenr*: generates the west-moving traffic arriving at the east side of the road segment. Traffic is generated based on a Poisson distribution with a rate of $\lambda = 1/10$.
- *oneWayTrafficRoad*: models the one-way traffic road segment, including the traffic control logic as well as the time for vehicles to cross the road segment. This model has two input ports *eastMovArival* and *westMovArrival*, which receive vehicles generated from *eastMovCarGenr* and *westMovCarGenr*, respectively. The vehicles finishing crossing the road segment are sent out as outputs through the *eastMovDeparture* and *westMovDeparture* output ports.

Since vehicles on both sides arrive at *oneWayTrafficRoad* randomly and the time it takes for a vehicle to cross the road segment is a random number, this model is a stochastic simulation model.

The *eastMovCarGenr* and *westMovCarGenr* atomic models are implemented as generator models similar to the one described in Section 3.5.3 (Pseudocode 3.1). For the *oneWayTrafficRoad* model, it uses several discrete phases and associated *sigma* (i.e., time advance) to process the incoming or queued vehicles on both directions. The following four phases and time advances (shown in the form of (phase, sigma)) are used:

- (*eastMovGreen_passive*, ∞): This is the phase when the east-moving traffic light is green, and there is no car moving on the road. In this phase, the road segment is waiting for cars to arrive.
- (*eastMovGreen_active*, *carPassingTime*): This is the phase when the east-moving traffic light is green, and there is a car moving on the road. The time advance *carPassingTime* is the time for the car to cross the road segment.
- (*westMovGreen_passive*, ∞): This is the phase when the west-moving traffic light is green, and there is no car moving on the road.
- (*westMovGreen_active*, *carPassingTime*): This is the phase when the west-moving traffic light is green, and there is a car moving on the road. The time advance *carPassingTime* is the time for the car to cross the road segment.

Besides the phase and sigma (denoted by x_{phase} and x_{sigma}, respectively) described above, the *oneWayTrafficRoad* model uses three other state variables: *elapsedTimeInGreen*, *eastQueue*, and *westQueue*, denoted by $x_{elapsedTimeInGreen}$, $x_{eastQueue}$, and $x_{westQueue}$, respectively. Together, the *oneWayTrafficRoad* model's state is defined by five state variables:

- x_{phase} defines the discrete phase of the road segment. This is a discrete categorical state variable as describe above.
- x_{sigma} is the time remaining in the current phase. This is a non-negative continuous state variable.
- $x_{elapsedTimeInGreen}$ keeps track of the elapsed time in the "eastMov-Green" or "westMovGreen" phases. This is a non-negative continuous state variable. *elapsedTimeInGreen* is reset to zero whenever the traffic light switches.
- $x_{eastQueue}$ keeps track of the size of the east-side queue. This is a discrete scalar state variable with non-negative integer values.
- $x_{westQueue}$ keeps track of the west-side queue. This is a discrete scalar state variable with non-negative integer values.

The *dataSavingTransducer* model is used to save data to be used in the data assimilation. This model is an extended version of the transducer model described in Section 3.5.3 (Pseudocode 3.3). This model saves two types of data for the *oneWayTrafficRoad* model. The first type is the input and output data related to the arrival and departure of vehicles in both directions. These data are used to generate the measurement data for the data assimilation. The second type is the state data of the *oneWayTrafficRoad* model. To obtain the state value at the time of each data assimilation step, *dataSavingTransducer* sends out a *requestData* message to the *oneWayTrafficRoad* model. After receiving this message, the *oneWayTrafficRoad* model updates its state variables accordingly. For example, *elapsedTimeInGreen* would be increased by the elapsed time from the last event. It then sends its state variables' values to the *dataSaving-Transducer* model through the *reportData* port.

Figure 7.10 shows the state trajectory of a sample simulation that runs for 1,000 s of simulation time. For better presentation, we do not show the specific phases (x_{phase}) but introduce a new variable called *trafficLightState* (denoted by $x_{trafficLightState}$) to represent the direction of the green traffic light. $x_{trafficLightState} =$

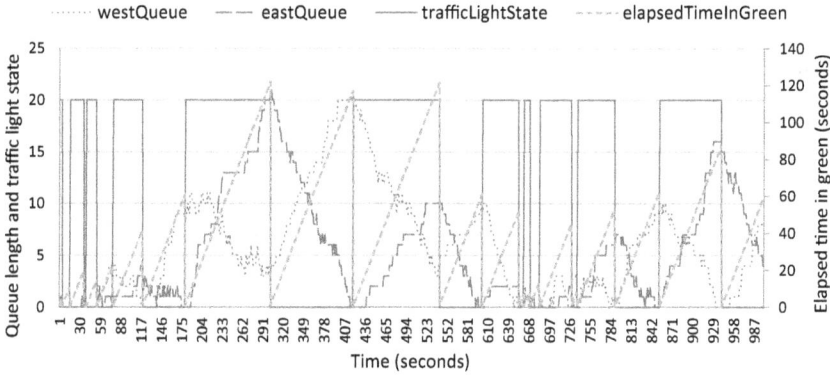

Figure 7.10.   Simulation results of the one-way traffic control system.

$eastMovGreen$ includes the phases of $eastMovGreen\_passive$ and $eastMovGreen\_active$ and is represented by a value of 20; $x_{trafficLightState} = westMovGreen$ includes the phases of $westMov$-$Green\_passive$ and $westMovGreen\_active$ and is represented by a value of 0. We do not include the x_{sigma} state variable in the figure either because x_{sigma} is the remaining time for specific phases and is not important for the following discussions. As a result, the four sate variables displayed in the figure are $x_{trafficLightState}$, $x_{elapsedTimeInGreen}$, $x_{eastQueue}$, and $x_{westQueue}$.

One can see that $x_{elapsedTimeInGreen}$ resets to zero every time $x_{trafficLightState}$ changes state and keeps increasing during the same traffic light state. The time it takes for the traffic light to switch direction is irregular, which is governed by the traffic-control rules and influenced by the number of vehicles waiting on the east- and west-side queues. When both queues have vehicles waiting, the traffic light switches after $x_{elapsedTimeInGreen}$ reaches 120 s. When the queues on both sides are small, the traffic light may switch frequently because once all the vehicles on one side pass the road segment, the traffic light switches immediately to allow vehicles on the other side to use the road. One can also see that when $x_{trafficLightState} = eastMovGreen$, $x_{westQueue}$ decreases because vehicles on the west side can move forward (note that the queue may also increase occasionally because new vehicles may arrive and join the queue). A similar pattern can be observed when $x_{trafficLightState} = westMovGreen$.

7.5.3 Experiment design

The goal of data assimilation is to dynamically estimate the state of the one-way traffic control system based on real-time measurement data collected by the observer. In this example, the measurement data are *eastMovDeparture* and *westMovArrival*. The system state incudes the five state variables of the *oneWayTrafficRoad* model described in the previous section. None of them can be directly derived from the measurement data.

We use the identical twin experiment design (Section 7.4) to set up the data assimilation experiments. For this example, the dynamic behavior is driven by the east-moving and west-moving traffic as well as the time for vehicles to cross the road segment, all of which are influenced by the random number generation in the model. By changing the seeds of random number generation, we can produce different simulation scenarios. Based on this idea, in the identical twin experiment, we set up the "true system" models using specific sets of random number seeds that are unknown to the data assimilation.

We consider two "true system" models, named *"true system" case I* and *"true system" case II*. The measurement data from the two "true system" cases are displayed in Figure 7.11. In both cases, the measurement data are collected every 30 s for a total period of 3,000 s.

To evaluate the data assimilation results, the state values at each data assimilation time from the two "true system" models are stored as the true state. Figure 7.12 shows the true state for the $x_{westQueue}$ and $x_{eastQueue}$ state variables for the two "true system" cases (the other state variables arc omitted here). One can see that the sizes of $x_{westQueue}$ and $x_{eastQueue}$ in both cases fluctuate over time, and there are significant differences between the two cases. For example, $x_{westQueue}$ in "true system" case I reaches high levels between 1,500 and 2,000 s, but the same state variable in "true system" case II stays relatively low until after 2,500 s. These results show that one cannot simply run a simulation (or multiple simulations) to reproduce the true states because different simulations generate significantly different state trajectories.

By examining the data shown in Figures 7.11 and 7.12, one can also see that while the measurement data provide some information about the true state, one cannot directly derive the true state values from the measurement data. Without knowing the specific

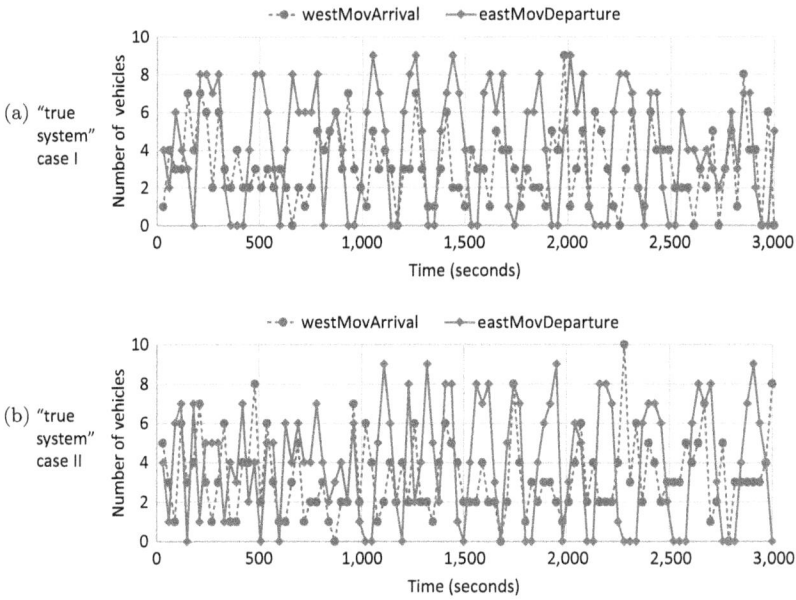

Figure 7.11.   Measurement data.

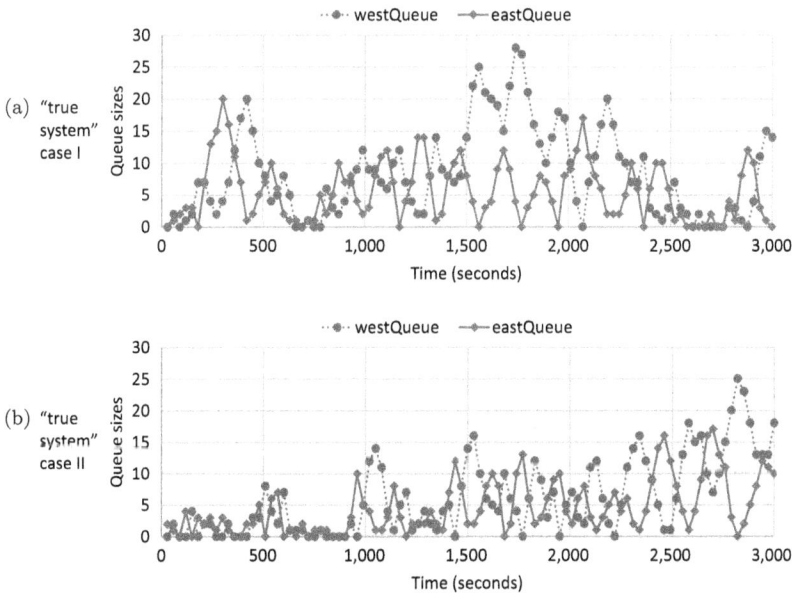

Figure 7.12.   The "true" state that needs to be estimated.

configurations of the "true system" models, the goal of data assimilation is to assimilate the measurement data to estimate the true state of the two cases.

7.5.4   *Particle-filter-based data assimilation*

This section describes the detailed setup for the particle-filter-based data assimilation.

Particle filter algorithm and parameters: The data assimilation algorithm is based on the bootstrap filter algorithm (Algorithm 6.4). It uses the following parameters:

- time step interval: 30 s;
- number of data assimilation steps: 100 steps;
- number of particles: 1,000 particles.

These parameters mean that the measurement data are collected every 30 s. The total time considered by the data assimilation is $30 \times 100 = 3,000$ s. The 1,000 number of particles is an arbitrary large number selected based on experience. The impact of this number is evaluated in Section 7.5.8.

State vector and state transition model: The data assimilation is to estimate the state of the *oneWayTrafficRoad* model. The following shows the state vector, which includes the five state variables described in Section 7.5.2:

$$x = \left(x_{phase}, x_{sigma}, x_{elapsedTimeInGreen}, x_{eastQueue}, x_{westQueue} \right)^T.$$
$$(7.1)$$

The state transition model is the simulation model described in Section 7.5.2. Both the *eastMovCarGenr* and *westMovCarGenr* models are essential parts of the simulation model because they model the traffic inputs of the *oneWayTrafficRoad* model.

To set up simulations for the state transition in particle filtering, the following configurations are needed. First, the initialize function of the *oneWayTrafficRoad* model can initialize the model based on the states defined by different particles. Second, each state transition runs a simulation for 30 s, which is the step interval for the data assimilation. Accordingly, the *dataSavingTransducer* model needs to be configured to request *oneWayTrafficRoad* to report its state at the end of the 30 s simulation so that the resulting state can be saved.

Particle initialization and particle rejuvenation: The particle filter algorithm starts from an initial set of particles, whose state variables are initialized in the following way:

- (x_{phase}, x_{sigma}): randomly chosen between $(eastMovGreen\_passive, \infty)$ and $(westMovGreen\_passive, \infty)$ with equal probability.
- $x_{elapsedTimeInGreen}$: randomly sampled between 0 and 120 s.
- $x_{eastQueue}$: sampled from the uniform distribution $U(0, 40)$ and then rounded to the closest integer. The number 40 is an arbitrary number that is sufficiently large for the queue length in this example.
- $x_{westQueue}$: sampled from a uniform distribution $U(0, 40)$ and then rounded to the closest integer.

To improve the robustness of data assimilation, we employ particle rejuvenation (see Section 7.3.3) to add perturbation noises to the particles before the sampling step. The perturbation noises for the five state variables are described as follows:

- (x_{phase}, x_{sigma}): No perturbation is introduced.
- $x_{elapsedTimeInGreen}$: Add a zero-mean truncated Gaussian noise that has a variance of σ^2 and lies within the range of $[-2\sigma, 2\sigma]$, with $\sigma = 2.0$.
- $x_{eastQueue}$: Add a zero-mean truncated Gaussian noise that has a variance of σ^2 and lies within the range of $[-5, 5]$, with $\sigma = 0.2x_{eastQueue}$ (i.e., 20% of the queue length).
- $x_{westQueue}$: a zero-mean truncated Gaussian noise is added in the same way as for $x_{eastQueue}$, with $\sigma = 0.2x_{westQueue}$.

In all the cases, if a resulting value after adding noise is negative, the value is set to 0. Furthermore, for $x_{eastQueue}$ and $x_{westQueue}$, the resulting values after adding noises are rounded to the closest integers.

Measurement data and weight computation: The measurement vector includes two variables:

$$y = (y_{eastMovDeparture}, y_{westMovArrival})^T, \qquad (7.2)$$

where $y_{eastMovDeparture}$ is the number of east-moving vehicles departing from the east side of the road segment and $y_{westMovArrival}$ is
the number of west-moving vehicles arriving at the east side of the
road segment. Both measurements are cumulative sums of the vehicle counts observed during the past time interval (i.e., 30 s). The
measurement data have noises, which are zero-mean Gaussian noises
sampled from $N(0, \sigma^2)$, where σ is set to be 10% of the actual number
of vehicles crossing the observer location. During a simulation, the
measurement data are recorded by the *dataSavingTransducer* model,
as described in Section 7.5.2. Examples of measurement data are
shown in Figure 7.11.

The two measurements are scalar Gaussian measurements and are
independent of each other. Thus, a particle's importance weight can
be computed using the likelihood probability computation based on
Equation (6.36). Let $y_k^{(i)} = (y_{eastMovDeparture,k}^{(i)}, y_{westMovArrival,k}^{(i)})^T$
be the predicted measurement for a particle $x_k^{(i)}$ at step k. This predicted measurement is collected by the *dataSavingTransducer* model
from the 30 s simulation corresponding to the ith particle's state
transition. Let $y_k^{real} = (y_{eastMovDeparture,k}^{real}, y_{westMovArrival,k}^{real})^T$ be the
real measurement data for step k. The real measurement is collected
by the *dataSavingTransducer* model from the simulation using the
"true system" model. Then, the importance weight $\bar{w}_k^{(i)}$ of particle
$x_k^{(i)}$ is computed as

$$\bar{w}_k^{(i)} = \bar{w}_{eastMovDeparture,k}^{(i)} \times \bar{w}_{westMovArrival,k}^{(i)}, \tag{7.3}$$

$$\bar{w}_{eastMovDeparture,k}^{(i)}$$

$$= \frac{1}{\sigma_w \sqrt{2\pi}} e^{-\frac{1}{2} \left(\frac{y_{eastMovArrival,k}^{(i)} - y_{eastMovArrival,k}^{real}}{\sigma_w} \right)^2}, \tag{7.4}$$

$$\bar{w}_{westMovArrival,k}^{(i)}$$

$$= \frac{1}{\sigma_w \sqrt{2\pi}} e^{-\frac{1}{2} \left(\frac{y_{westMovArrival,k}^{(i)} - y_{westMovArrival,k}^{real}}{\sigma_w} \right)^2}, \tag{7.5}$$

where σ_w is set to be 0.6 in this example.

7.5.5 *Dynamic state estimation*

We first focus on the problem of dynamic state estimation. Figure 7.13 shows the data assimilation results for estimating $x_{trafficLightState}$ and $x_{elapsedTimeInGreen}$ for "true system" case I. The state estimate in each step is computed as the average of all the particles (after resampling). Since $x_{trafficLightState}$ is a categorical state, the average of $x_{trafficLightState}$ is computed based on a numerical value of 20 if $x_{trafficLightState} = eastMovGreen$ and 0 if $x_{trafficLightState} = westMovGreen$ (see Figure 7.10). For comparison purposes, the figure also displays the corresponding true state from "true system" case I. From the figure, one can see that the state estimates for $x_{trafficLightState}$ and $x_{elapsedTimeInGreen}$ match the corresponding "true" states well in most steps.

The averaged result from all the particles in Figure 7.13 is a simplified way of presenting the data assimilation result. It does not give the full details of the distribution of particles' estimates. To demonstrate the detailed result from all the particles, we select a representative step: step 67 (step time is 2010 s) and examine all

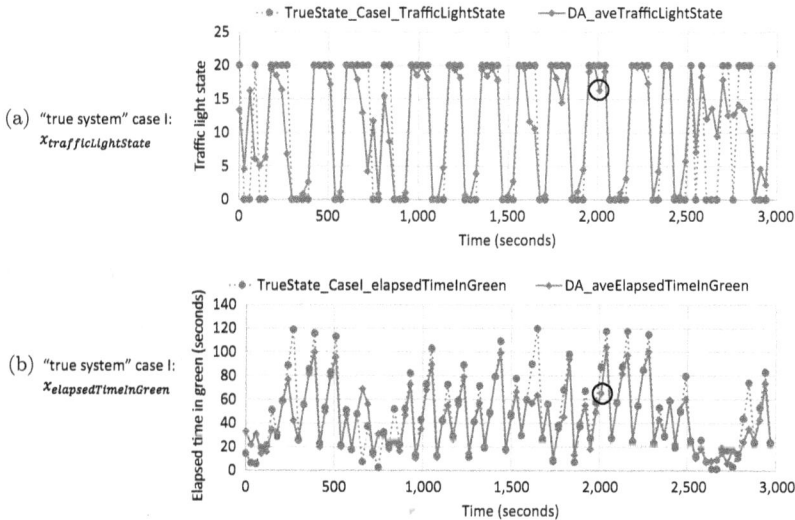

(a) "true system" case I: $x_{trafficLightState}$

(b) "true system" case I: $x_{elapsedTimeInGreen}$

Figure 7.13. State estimation result for $x_{trafficLightState}$ and $x_{elapsed}$ $TimeInGreen$. In the figure, the "true" state values are denoted with the prefix "TrueState_"; the results from data assimilation are denoted with the prefix "DA_".

the particles in detail for this step. For step 67, the averaged state estimates for $x_{trafficLightState}$ and $x_{elapsedTimeInGreen}$ are circled in Figure 7.13. The true $x_{trafficLightState}$ is $eastMovGreen$ (value of 20) and the true $x_{elapsedTimeInGreen}$ is 87.5 s. For this step, among the 1,000 particles, 817 particles' $x_{trafficLightState}$ is $eastMovGreen$ and 183 particles' $x_{trafficLightState}$ is $westMovGreen$. The histograms of $x_{elapsedTimeInGreen}$ for the 817 particles whose $x_{trafficLightState}$ is $eastMovGreen$ and for the 183 particles whose $x_{trafficLightState}$ is $westMovGreen$ are shown in Figure 6.4 in Chapter 6. Figure 6.4 showed that the belief of $x_{elapsedTimeInGreen}$ is bimodal: one concentrates in the range of [70, 93] s; the other concentrates in the range of [1, 15] s. The $x_{elapsedTimeInGreen}$ estimates from the majority particles are close to the true value of 87.5 s.

Figure 7.13 shows that the $x_{trafficLightState}$ estimates are less accurate during the time periods when the "true system" traffic light switches frequently (e.g., during 0–200 s and after 2,500 s). This is understandable because the frequent changes of traffic light during those time periods result in measurement data that carry less distinct information about the traffic light state.

Similar results for $x_{trafficLightState}$ and $x_{elapsedTimeInGreen}$ are obtained for "true system" case II. In the following description, we omit the results for $x_{trafficLightState}$ and $x_{elapsedTimeInGreen}$, and focus only on the estimation results for $x_{eastQueue}$ and $x_{westQueue}$ due to their importance in influencing the traffic delay time.

Figure 7.14 shows the data assimilation results for estimating $x_{westQueue}$ and $x_{eastQueue}$ for both "true system" case I and case II. As before, the state estimates in each step are computed as the average of all the particles. The corresponding "true" states from the two "true system" models are also displayed in the figure.

One can see that in both cases, after the first several steps, the state estimates for $x_{westQueue}$ and $x_{eastQueue}$ begin to track the corresponding "true" states. Take case II as an example: Before 1,000 s when the true $x_{westQueue}$ and $x_{eastQueue}$ are small, the state estimates are also small; after 2,000 s when the true $x_{westQueue}$ and $x_{eastQueue}$ become large, the state estimates become large too. Furthermore, the cyclic increase and decrease of the estimated $x_{westQueue}$ and $x_{eastQueue}$ generally match those of the corresponding true states. $x_{eastQueue}$ has smaller estimation errors because the measurement data $y_{westMovArrival}$ is a direct observation of the number of vehicles

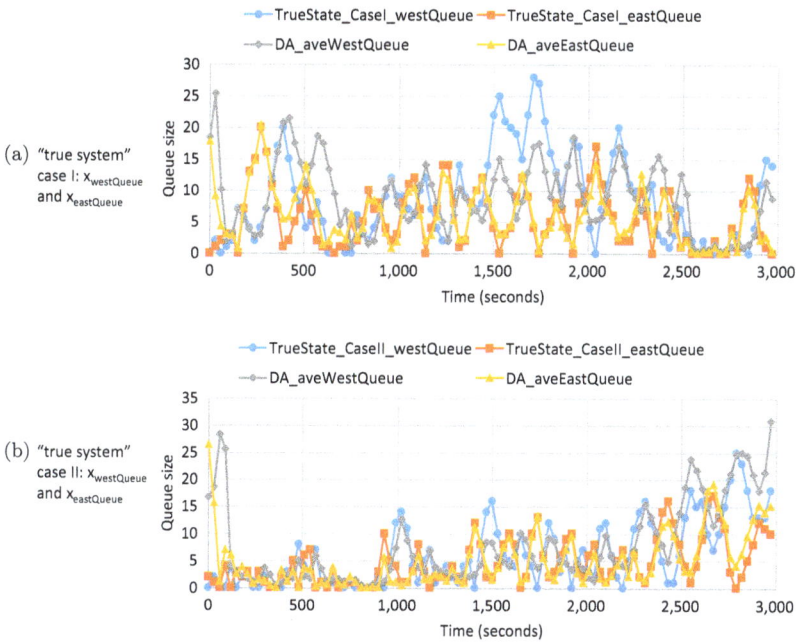

(a) "true system" case I: $x_{westQueue}$ and $x_{eastQueue}$

(b) "true system" case II: $x_{westQueue}$ and $x_{eastQueue}$

Figure 7.14. State estimation results for $x_{westQueue}$ and $x_{eastQueue}$.

arriving on the east side of the road segment. Although there is no measurement data on the west side, the data assimilation is able to estimate the true $x_{westQueue}$ in most of the steps with relatively small errors. This is because the data assimilation uses information from both the simulation model and measurement data.

Quantitative results measuring the accuracy of state estimation will be provided in Section 7.5.8. In this experiment, the measurement data includes two measurement data on the east side: $y_{eastMovDeparture}$ and $y_{westMovArrival}$. The impact of using less measurement data (by removing $y_{westMovArrival}$) and more data (by adding new measurement data on the west side) will be shown in Section 7.5.8 too.

7.5.6 *Online model calibration*

The dynamic state estimation described above assumes all the parameters of the simulation model are predefined and known. This section demonstrates online model calibration, where one or more parameters of the simulation model are unknown and need to be estimated based on measurement data. We consider the following two scenarios of online model calibration: (1) static parameter estimation; and (2) dynamic parameter estimation. The former estimates a static parameter that does not change over time. The latter estimates a parameter that dynamically changes over time.

An important parameter of the one-way traffic-control system is the average time it takes for vehicles to move across the road segment, denoted by $\theta_{crossingTime}$. In the default model described above, the crossing time is modeled by a truncated normal distribution that has a mean of $\mu = \theta_{crossingTime}$ and a variance of $\sigma^2 = 0.5^2$ and lies within the range of $[\mu - \Delta, \mu + \Delta]$, where $\theta_{crossingTime} = 4.0\,\text{s}$ and $\Delta = 1.0\,\text{s}$. In this section, we assume that the crossing time distribution follows the same truncated normal distribution described above; however, the mean $\mu = \theta_{crossingTime} = 4.0$ is unknown and needs to be estimated.

In the static parameter estimation scenario, $\theta_{crossingTime}$ is a constant and needs to be estimated. To set up the identical twin experiment for this scenario, the "true system" model is configured in the same way as in "true system" case I described in Section 7.5.3, which has a constant $\theta_{crossingTime} = 4.0$. The measurement data are the same as shown in Figure 7.11(a). However, $\theta_{crossingTime}$ of the simulation model used in data assimilation does not have a predefined value. We use data assimilation to carry out joint state–parameter estimation to estimate this static parameter together with the state of the "true system."

To carry out joint state–parameter estimation, the state vector is expanded to include the parameter $\theta_{crossingTime}$ that needs to be estimated, as shown in Equation (7.6). The expanded state vector means that each particle represents not only the state variables but also the model parameter $\theta_{crossingTime}$. The state variables are used for starting the state transition in each data assimilation step, and the model parameter is used for parametrizing the simulation model governing the state transition:

$$x = (x_{phase}, x_{sigma}, x_{elapsedTimeInGreen}, x_{eastQueue} x_{westQueue},$$

$$\times \; \theta_{crossingTime})^T. \tag{7.6}$$

To set up the data assimilation, the particle initialization and particle rejuvenation treat the first five state variables in the same

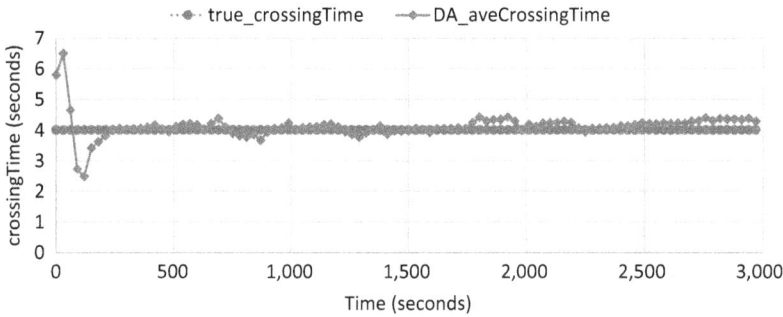

Figure 7.15.   Joint state–parameter estimation for a static parameter.

way as described in Section 7.5.4. The new variable $\theta_{crossingTime}$ is a continuous variable. In particle initialization, $\theta_{crossingTime}$ is sampled from the uniform distribution $\sim U(2.0, 10.0)$. To allow the estimated parameter values to change and converge to the true parameter value, the dynamics of the model parameter is modeled as a zero-mean Gaussian random walk. Specifically, in each step's particle rejuvenation, a particle adds a zero-mean Gaussian perturbation to its estimated $\theta_{crossingTime}$. Since the true $\theta_{crossingTime}$ is a static parameter, the Gaussian perturbation has a relatively small variance ($\sigma^2 = 0.1^2$). The perturbated $\theta_{crossingTime}$ is truncated to make it fall within the range of $[2.0, 10.0]$.

The particle filter algorithm uses 3,000 particles. The increased number of particles is due to the fact that the state vector includes an extra variable that needs to be estimated. Figure 7.15 shows the estimated parameter values (denoted by *DA_aveCrossingTime*) in comparison with the true parameter values (denoted by *true_crossingTime*) in each step of the data assimilation. The estimated parameter values are averaged over all the particles. The figure shows only the parameter estimation results. The state estimation results are very similar to what is shown in Figures 7.13 and 7.14(a) and are omitted here. One can see that the true $\theta_{crossingTime}$ is 4.0 all the time. After four steps of the data assimilation, the estimated $\theta_{crossingTime}$ starts to correctly converge to the true value and maintains around the true value. This experiment shows that the data assimilation is able to jointly estimate the unknow static parameter together with the system state.

We next consider dynamic parameter estimation. To set up the identical twin experiment for this scenario, the "true system" model is configured in the same way as in "true system" case I described in Section 7.5.3, except that $\theta_{crossingTime}$ is not a constant any more — it changes dynamically in the following way:

- Between 0 and 600 s, $\theta_{crossingTime}$ maintains to be 4.0.
- Between 600 and 1,000 s, $\theta_{crossingTime}$ linearly increases from 4.0 to 6.0, with a 0.005/s increasing rate.
- Between 1,000 and 1,800 s, $\theta_{crossingTime}$ linearly decreases from 6.0 to 2.0, with a −0.005/s decreasing rate.
- $\theta_{crossingTime}$ goes back to 4.0 at 1,800 s and maintains at 4.0 afterward.

The data assimilation is set up in the same way as that in the static parameter scenario described above, except that the particle rejuvenation for $\theta_{crossingTime}$ uses a Gaussian perturbation that has a larger variance ($\sigma^2 = 0.3^2$). A larger perturbation is used because $\theta_{crossingTime}$ dynamically changes over time as opposed to being a constant.

Figure 7.16 shows the data assimilation results for the dynamic parameter estimation scenario. Figure 7.16(a) shows the estimation results for $\theta_{crossingTime}$; Figure 7.16(b) shows the estimation results for the two state variables $x_{westQueue}$ and $x_{eastQueue}$. As before, the data assimilation results are averaged over all the particles.

From Figure 7.16(a), one can see that the estimated $\theta_{crossingTime}$ is able to track the dynamic changes in the true $\theta_{crossingTime}$ over time. For example, during the time when the true $\theta_{crossingTime}$ increases from 4.0 to 6.0 between 600 and 1,000 s and then decreases from 6.0 to 2.0 between 1,000 and 1,800 s, the estimated $\theta_{crossingTime}$ increases and then deceases too. During the time when the true $\theta_{crossingTime}$ maintains at 4.0 (before 600 s and after 1,800 s), the estimated $\theta_{crossingTime}$ converges and stays at around 4.0 too. Note that compared to the static parameter case (Figure 7.15), even during the time when the true $\theta_{crossingTime}$ is 4.0, the estimated $\theta_{crossingTime}$ has larger variations because a larger perturbation is used during particle rejuvenation. Figure 7.16(a) also shows that the estimated $\theta_{crossingTime}$ lags behind the changes in the true $\theta_{crossingTime}$.

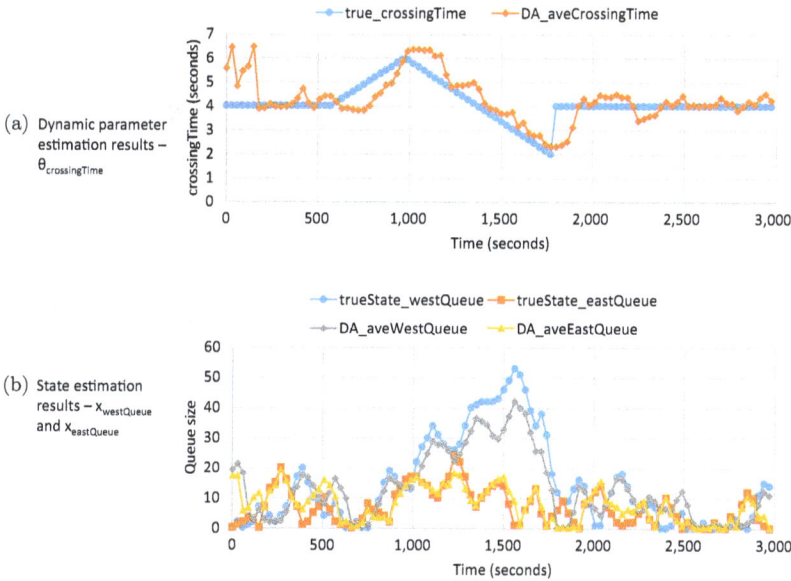

Figure 7.16.   Joint state–parameter estimation for a dynamic parameter.

This is because, without knowing if the true $\theta_{crossingTime}$ actually changes or not, the particles tend to maintain their $\theta_{crossingTime}$ values unless there is significant evidence from measurement data showing otherwise. This tendency results in a delay for the estimated $\theta_{crossingTime}$ to track the changes of the true $\theta_{crossingTime}$.

Figure 7.16(b) shows that the estimated $x_{westQueue}$ and $x_{eastQueue}$ are able to track the changes in the true $x_{westQueue}$ and $x_{eastQueue}$. In particular, due to the increase in $\theta_{crossingTime}$ between 600 and 1,000 s in the "true system" model, the true $x_{westQueue}$ significantly increases after 600 s and stays at high levels until 1,800 s ($x_{eastQueue}$ increases too, but its increase is less significant). Should the data assimilation still use a constant $\theta_{crossingTime}$ (e.g., $\theta_{crossingTime} = 4.0$) in its simulation model, it would be difficult for the data assimilation to correctly estimate the significant increase in $x_{westQueue}$. By treating $\theta_{crossingTime}$ as a dynamic parameter and correctly estimating its value using joint state–parameter estimation, the data assimilation is able to more accurately track the true $x_{westQueue}$ and $x_{eastQueue}$.

7.5.7 *Simulation-based prediction/analysis*

The data assimilation results presented above allow us to carry out simulation-based prediction/analysis for the one-way traffic control system. Examples of prediction/analysis for this system include predicting the real-time traffic delay time (described in the following) or analyzing whether changing the traffic-control rules (e.g., making the traffic light switches more or less frequent) would reduce the overall traffic delay time. A simulation-based prediction/analysis would need to use the real-time state estimates to create simulations to predict/analyze the system's future behavior. When model parameters are estimated, the simulation model would be parameterized using the estimated model parameters.

To demonstrate simulation-based prediction/analysis, this section considers the example of predicting the *real-time traffic delay time*. The real-time traffic delay time at time t is denoted by D_t and is defined as the time for a newly arrived vehicle at time t to finish crossing the road segment. This includes the time for the vehicle to wait in the queue and to move through the road segment when the traffic light is green. D_t changes dynamically, influenced by the number of vehicles waiting in the queue and the traffic light state at time t as well as how the one-way traffic-control system will work after time t. Due to the asymmetric traffic on the two sides of the road segment, the real-time delay time for east-moving and west-moving vehicles would be different. We denote the east-moving delay time by D_t^{east} and the west-moving delay time by D_t^{west}.

To predict D_t^{east} and D_t^{west} at the time of each data assimilation step, after each data assimilation, we run simulations to simulate the dynamics of the one-way traffic control system, starting from the "current" state estimated from the data assimilation. The simulations need to take into consideration the future vehicles that will arrive at the road segment too because the one-way traffic-control rules are dynamically influenced by the number of vehicles arriving at both sides of the road segment. Since the estimated state is represented by a set of particles, a simulation is created for each particle with the initial state defined by the particle's value. More specifically, at time t_k of step k, each particle in the resampled particle set would create a simulation including the *oneWayTrafficRoad* model as well as the *eastMovCarGenr* and *westMovCarGenr* models shown in Figure 7.9. Let $x_k^{(i)}$, $i = 1, 2, \ldots, N$ be the set of particles at step k. For each particle $x_k^{(i)}$, the *one Way Traffic Road* model is initialized using the particle state $x_k^{(i)}$. To simulate the delay time for vehicles arriving at t_k, we add a special vehicle at the ends of the west-side and east-side queues during the initialization of the simulation model. The time it takes for the two special vehicles to finish crossing the road segment is monitored and used as the predicted east-moving and west-moving delay times from the corresponding particle. The prediction results from all the particles then represent the distribution of the predicted results.

Figure 7.17 shows the predicted D_t^{east} and D_t^{west} at each data assimilation step for "true system" case II described in Section 7.5.3 (prediction results of "true system" case I can be found in Hu (2022), and is omitted here). The predicted delay time is computed as the average based on the predictions from all the particles. In the figure, the horizontal axis is the vehicle arriving time and the vertical axis is the traffic delay time. A data point (t_x, t_y) means that at time t_x, the predicted (or actual) delay time is t_y. Since the predictions happen only at the data assimilation steps, the predicted delay time has data points only every 30 s. To evaluate the prediction results, the actual east-moving and west-moving delay times for each vehicle in the "true system" model is measured and displayed in the figure too. Note that for the actual delay time data points, their t_x values do not match exactly with those from the predicted delay time data points. This is because the "true system" model may not have a vehicle arriving at exactly the data assimilation time.

Figure 7.17 shows that the predicted delay time matches the overall trend of the true traffic delay time. The predictions have large errors during the first five data assimilation steps. This is understandable because during those steps, the data assimilation results have not converged. In the following analysis, we exclude those steps. Table 7.2 shows a quantitative analysis of the prediction results for both "true system" case I and case II. We define the *absolute error (AE)* of prediction in each data assimilation step as the absolute difference between the actual delay time and the predicted delay time. Table 7.2 shows the *mean absolute error (MAE)*, which is the mean AE from all the steps, and the percentage of steps whose AE is less than 30.0 s. One can see that except for the predicted D_t^{east} of case

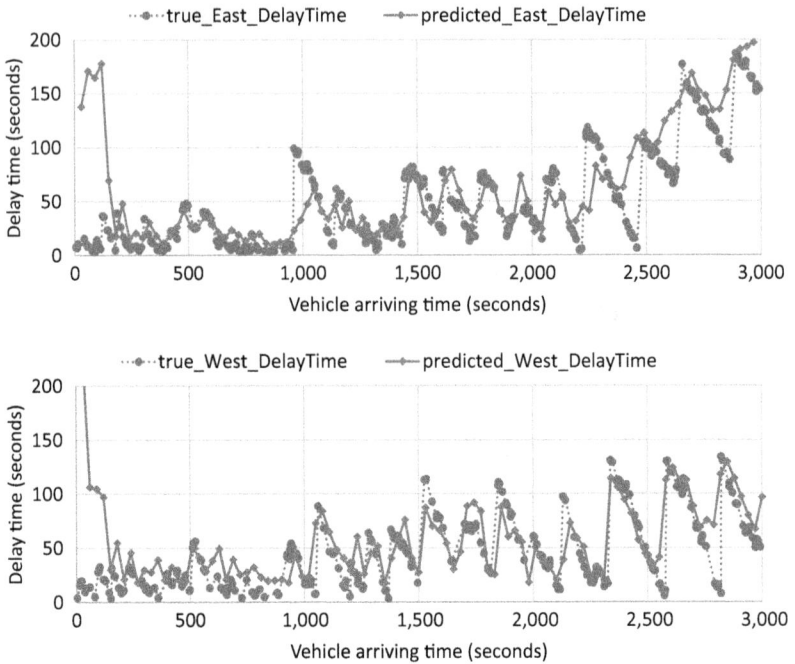

Figure 7.17. Simulation-based prediction of real-time traffic delay time for "true system" case II.

Table 7.2. Simulation-based prediction results.

	Case I		Case II	
	D_t^{east}	D_t^{west}	D_t^{east}	D_t^{west}
MAE (s)	25.7	19.6	17.5	17.1
Percentage of steps with AE <30	72.9%	83.3%	84.2%	87.3%

I that has an MAE of 25.7 s, all other predictions' MAE are less than 20 s. Furthermore, in all the predictions except for the D_t^{east} of case I, more than 80% of the steps has AE less than 30 s. Compared to the D_t^{east}, the D_t^{west} is more accurate because the data assimilations provide more accurate estimations for $x_{eastQueue}$, as discussed in Section 7.5.5.

The above results are based on the data assimilations described in Section 7.5.5, which assume that the model parameters are known.

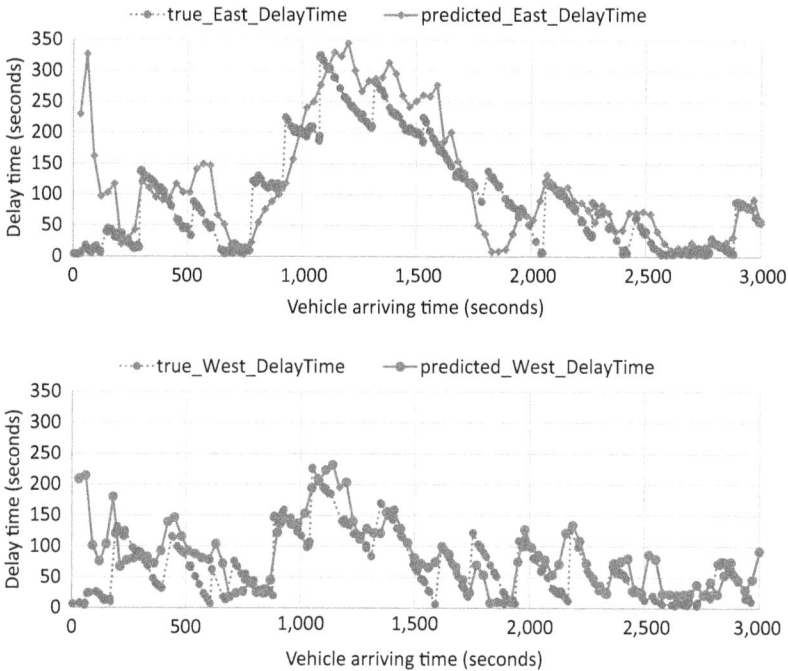

Figure 7.18. Simulation-based prediction of real-time traffic delay time based on joint state–parameter estimation.

When the model parameters are unknown, the estimated parameter values would be used to parameterize the simulation model for simulation-based prediction. Figure 7.18 shows the predicted traffic delay time for the experiment of online model calibration that estimates the dynamically changing model parameter described in Section 7.5.6 (Figure 7.16).

One can see that the predicted traffic delay time using the joint state–parameter estimation is able to match the actual traffic delay time well. In particular, the simulation-based prediction is able to accurately predict the long delays for the east-moving vehicles in the time range of 800–1800 s. This is because the data assimilation is able to estimate the large size of the west-side queue during that time period (see Figure 7.16). These prediction results can provide valuable information for real-time decision-making for traffic control (e.g., to route traffic away from the construction road segment).

7.5.8 *Data assimilation performance*

Several factors influence the data assimilation performance. These include the number of particles used by the particle filter algorithm and the availability and quality of measurement data. Generally, using more particles would improve the data assimilation results; nevertheless, more particles also mean a higher computation cost. Similarly, more measurement data and higher-quality data would lead to more accurate data assimilation results. This section demonstrates the impact of these factors on the data assimilation performance based on the tutorial example described above. It considers both the accuracy and execution time of data assimilation.

We use the root mean square error (RMSE) to quantify the data assimilation accuracy, which is calculated based on the differences between the estimated state values and the corresponding true state values. For the dynamic state estimation problem, the state vector includes five state variables: x_{phase}, x_{sigma}, $x_{elapsedTimeInGreen}$, $x_{eastQueue}$, and $x_{westQueue}$. For simplicity, we consider only $x_{eastQueue}$ and $x_{westQueue}$ when computing the RMSE. Since the data assimilation results are represented by a set of particles, we calculate the RMSE using the estimations from all the particles. In each data assimilation step, an RMSE is calculated by averaging all particles' RMSEs computed based on $x_{eastQueue}$ and $x_{westQueue}$. Then, an overall RMSE for a data assimilation run is defined as the average of the RMSEs from all the data assimilation steps. Due to the stochastic nature of data assimilation, for each set of parameters, we run data assimilation 20 times (each one is referred to as a *data assimilation run*) and compute the average RMSE (denoted by $RMSE^{ave}$) as the accuracy measurement. The standard deviation of the RMSE (denoted by $RMSE^{stdv}$) from the 20 runs is also computed, which is an indication of the consistency of the data assimilation results from the data assimilation runs.

Number of particles: We first show the impact of the particle number on the state estimation accuracy. Tables 7.3 and 7.4 show the accuracy results under different numbers of particles for the dynamic state estimation "true system" case I and case II, respectively, described in Section 7.5.5.

The results show that when the particle number is 400, the $RMSE^{ave}$ and $RMSE^{stdv}$ for both cases are the largest. This means when the number of particles is small, there are relatively large errors as well as significant variations in the state estimation results from the different data assimilation runs. In other words, the small number of particles cannot produce robust data assimilation results. As the particle number increases, $RMSE^{ave}$ and $RMSE^{stdv}$ decrease. After the particle number reaches 1,000, increasing the number further does not significantly reduce $RMSE^{ave}$ and $RMSE^{stdv}$ anymore. This means that 1,000 particles are sufficient to produce robust data assimilation results for this example.

It is important to note that the number of particles needed for producing robust results would be different for different applications. In general, more particles would be needed for estimating state vectors that have more state variables.

Number of measurement data: All the experiments described so far are based on a default measurement setting that includes two measurement variables: $y_{eastMovDeparture}$ and $y_{westMovArrival}$, both of which are observed from the east side of the road segment. We name this setting as the *Two Measurement* setting. To evaluate the impact of the number of measurement data on data assimilation, we consider two other measurement settings, named as the *One Measurement* and *Four Measurement* settings. These measurement settings are described as follows:

Table 7.3. Impact of particle number on estimation accuracy for "true system" case I.

Number of particles	400	600	800	1,000	1,500	2,000
$RMSE^{ave}$	7.1	6.5	6.0	5.9	6.0	5.0
$RMSE^{stdv}$	3.5	1.9	1.1	0.6	0.8	0.6

Table 7.4. Impact of particle number on estimation accuracy for "true system" case II.

Number of particles	400	600	800	1,000	1,500	2,000
$RMSE^{ave}$	5.6	5.5	4.9	4.4	4.6	4.6
$RMSE^{stdv}$	2.0	1.6	1.2	0.4	0.4	0.3

Table 7.5.   Impact of measurement data on data assimilation accuracy.

Measurement setting	One measurement	Two measurement	Four measurement
"true system" case I $RMSE^{ave}$	6.9	5.9	4.4
"true system" case II $RMSE^{ave}$	5.2	4.4	3.0

- **One Measurement:** The measurement data include only one variable $y_{eastMovDeparture}$.
- **Two Measurement:** The measurement data include two variables: $y_{eastMovDeparture}$, $y_{westMovArrival}$. This was the default setting used in all the previous experiments.
- **Four Measurement:** The measurement data include four variables: $y_{eastMovDeparture}$, $y_{westMovArrival}$, $y_{westMovDeparture}$, and $y_{eastMovArrival}$. In other words, the measurement data also include observations from the west side of the road segment.

For the One Measurement setting, the weight computation is based on Equation (7.4) only. For the Four Measurement setting, the weight computation is an extension of Equation (7.3) to include contributions from measurement data $y_{westMovDeparture}$ and $y_{eastMovArrival}$ too. Table 7.5 shows the impact of the three measurement settings on $RMSE^{ave}$ for "true system" case I and case II. All the data assimilation runs in Table 7.5 use 1,000 particles.

Two observations can be made from Table 7.5. First, the One Measurement setting that uses only $y_{eastMovDeparture}$ is still able to generate relatively good data assimilation results. This is because $y_{eastMovDeparture}$ carries information about how many vehicles crosses the one-way traffic-control road. This information reflects the traffic light state as well as the west-side and east-side queue sizes and thus allowing the data assimilation to estimate those states. As a comparison, the measurement data $y_{westMovArrival}$ carries information only about the number of vehicles arriving at the east side of the road segment; it does not reflect how the one-way traffic road works. Thus, if the One Measurement setting was based on $y_{westMovArrival}$, it would lead to poor data assimilation result. This means not every measurement data carry the same amount of information from the data assimilation point of view.

Second, Table 7.5 shows that when more measurement data are used, the data assimilation generates more accurate state estimation results (as indicated by the smaller $RMSE^{ave}$). This is expected because compared to the One Measurement setting, the Two Measurement setting has more information about the number of vehicles arriving at the east side of the road segment. This extra information allows the data assimilation to achieve more accurate state estimates, especially for the estimation of the east-side queue. Similarly, compared to the Two Measurement setting, the Four Measurement setting has extra information about the west side of the road segment, thus allowing the state estimation accuracy to be improved further.

In summary, more measurement data help improve the data assimilation results. Furthermore, not all measurement data carry the same amount of information. When there are limited sensor resources, how to deploy them to collect the most useful information for data assimilation is an interesting research topic.

Execution time of particle-filter-based data assimilation: The execution time of data assimilation is an important factor to consider when carrying out data assimilation for real applications. In a real-time context, where measurement data arrive sequentially, it is crucial to finish the data assimilation computation within the time interval of a data assimilation step. Compared to Kalman-filter-based data assimilation, data assimilation based on particle filters has a higher computation cost due to the large number of particles that are often required. This is especially true if working with discrete simulations because each particle needs to invoke a simulation as part of the sampling step.

To provide an example of the execution time of particle-filter-based data assimilation, Table 7.6 shows the execution time of the data assimilation for the tutorial example when using different numbers of particles. The data assimilation considered in Table 7.6 is based on the dynamic state estimation for "true system" case I. For each particle number, we measure the execution time from 20 data assimilation runs and compute the average execution time (denoted by $ExecutionT^{ave}_{DA\_run}$). Since each data assimilation run has 100 steps, the average execution time for each data assimilation step (denoted by $ExecutionT^{ave}_{DA\_step}$) is also calculated by dividing $ExecutionT^{ave}_{DA\_run}$ by 100. All the experiments are run on a Dell

Table 7.6.   Execution time of particle-filter-based data assimilation.

Number of particles	400	600	800	1,000	1,500	2,000
$ExecutionT^{ave}_{DA\_run}$ (seconds)	5.4	7.8	10.4	13.1	18.4	24.2
$ExecutionT^{ave}_{DA\_step}$ (seconds)	0.054	0.078	0.104	0.131	0.184	0.242

laptop that has an Intel i7-9750H CPU with a 2.60 GHz processor, 32.0 GB of RAM, and Windows 10 Pro OS.

Table 7.6 shows that $ExecutionT^{ave}_{DA\_run}$ linearly increases as the number of particles increases. This is expected because the computation complexity of the data assimilation algorithm is directly related to the number of particles. In the case of 1,000 particles, it takes 13.1 s to finish a data assimilation run. The average execution time for each data assimilation step is 0.131 s. For this tutorial example, the time interval of each data assimilation is 30 s. The 0.131 s computation time means that the data assimilation computation can be easily finished before the next data assimilation step starts.

In general, the execution time of particle-filter-based data assimilation would depend on the number of particles as well as the complexity of the simulation model, both of which are application specific. More complex applications typically have higher computation costs due to the complexity of the simulation model as well as the larger number of particles that are needed.

7.6   Sources

Some of the descriptions in Sections 7.3 and 7.4 are adapted from the work of Hu and Wu (2019). Some results related to the tutorial example have been published in the work of Hu (2022).

Part 3

Application and Look Ahead

Chapter 8

The Wildfire Spread
Simulation Application

8.1 Introduction

Previous chapters focus on the concepts and principles and use demonstrative examples that are intentionally kept simple. This chapter describes how dynamic data-driven simulation (DDDS) can be applied to a more complex application: the wildfire spread simulation application. Wildfire spread is an example of a *spatiotemporal system*, whose states change in both space and time. Many other spatiotemporal systems exist, including road traffic, pedestrian crowds, and infectious disease spread. There is growing interest in assimilating real-time data into the simulations of these systems to predict and analyze their spatiotemporal behaviors.

Spatiotemporal systems bring about several complexities to data assimilation and DDDS. First, spatiotemporal systems often have many state variables in order to adequately capture the system states that are spatially dependent. For example, a large wildfire would have a large number of location-based state variables to define the (dynamically changing) shape of its fire front. The many state variables result in a high-dimensional state, which makes it more difficult for data assimilation to converge to the true state. Second, spatiotemporal systems have measurement data that are spatially dependent, i.e., reflecting information only for the local areas. This brings about challenges in effectively assimilating localized measurement data for estimating the full system state. Last but not least, spatiotemporal systems typically involve complex simulation models that have high computation costs. Efficient computation of the simulation and data assimilation is essential in order to meet the real-time requirements of DDDS.

The material in this chapter is derived from the author's previous research on data assimilation for wildfire spread simulation. This chapter complements the previous chapters by using the wildfire spread simulation application to demonstrate how advanced data assimilation can be carried out to support DDDS for large-scale spatiotemporal systems.

8.2   DDDS for Wildfire Spread Simulation

One of the most salient features of a wildfire is that its fire front evolves in space and time. The spatiotemporal fire spread behavior is heterogenous, meaning that some portions of the fire front spread faster than others, and the spread speeds vary over time. Figure 8.1 illustrates wildfires' spatiotemporal spread behavior using the day-to-day progression of the Station Fire as an example.

The three important factors influencing wildfire spread are vegetation, terrain, and weather (Pyne *et al.*, 1996). In the wildfire literature, vegetation is described by fuels, which refer to the composite of variables that describe the vegetation and other combustible biomass the fire spreads through. Terrain information includes slope and aspect, where slope is the inclination of a land surface and aspect is the direction the surface is facing. Compared to fuel and terrain, weather has a much more dynamic influence on wildfire behavior. The three components of weather that greatly influence fire spread are wind speed, wind direction, and atmospheric moisture content. In addition, air temperature can also have significant dynamic effect — initial fire risk is higher at higher temperatures, while fire-induced heating and buoyancy can drive fire storms. Based on these factors, fire spread can be described as the propagation of a flaming fire front that involves a series of ignitions whose heat brings successive stripes of fuel to the ignition temperature via a contagion process. This process is considered to be in a steady state for homogeneous fuels and steady weather conditions but an unsteady state for inhomogeneous fuels and dynamic weather (Rothermel, 1972).

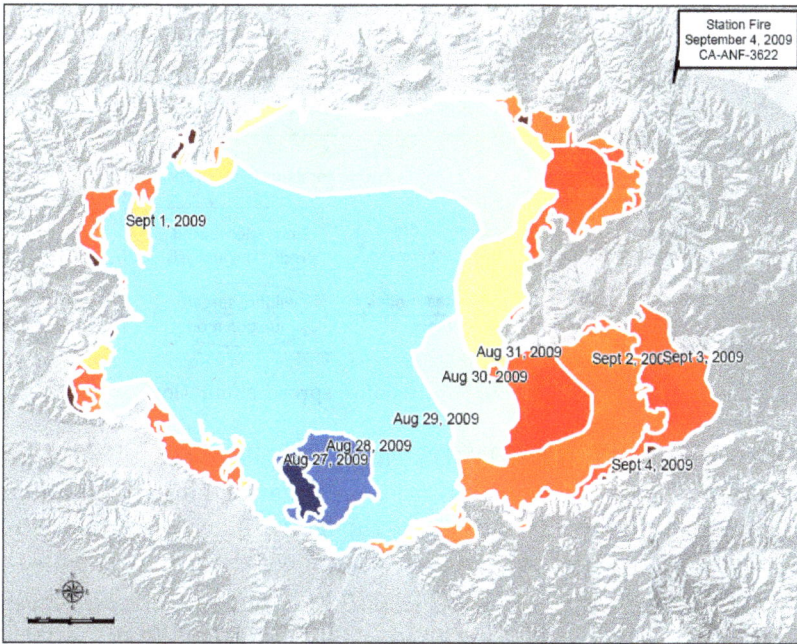

Figure 8.1. Spatiotemporal fire spread behavior exemplified by the day-to-day progression of the Station Fire from August 27, 2019 to September 04, 2019. The Station fire was a wildfire that happened in Los Angeles County, California, USA, in 2009 that burned 650 km^2 in total. The figure shows that the fire spread extremely fast on the day of August 29, 2009, growing in all directions but had the largest growth on the north side.

Source: the InciWeb (https://inciweb.nwcg.gov/).

Simulations have long been used to study the spatiotemporal behavior of wildfire spread. These simulations provide useful information for fire managers to understand fire behavior and to analyze wildfire risk. Nevertheless, to this day, they have played limited roles in serving as operational tools for real-time wildfire spread prediction and decision-making. Accurate simulation of wildfire spread remains a challenging task due to the dynamic nature of fire behavior, deficiencies of existing simulation models, and difficulties in obtaining accurate fuel and weather data. Assimilating real-time measurement data from wildfires can alleviate these limitations and bring wildfire spread simulation one step closer to providing real-time decision support.

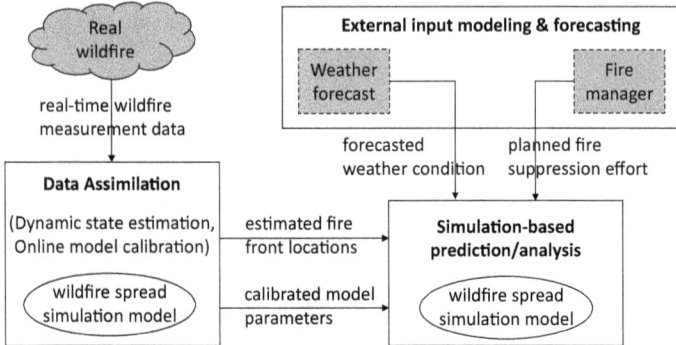

Figure 8.2.　DDDS for wildfire spread simulation.

Figure 8.2 shows how DDDS can be applied to wildfire spread simulation. The four DDDS activities are organized into three components represented by three boxes in the figure. The *data assimilation* component assimilates real-time measurement data from a real wildfire to provide *dynamic state estimation* of fire front locations and *online model calibration*. The data assimilation results are used by the *simulation-based prediction/analysis* component to predict future fire spread, which also takes inputs from the *external input modeling & forecasting* component regarding future weather condition and planned fire suppression effort. In the following, we elaborate on the components, model, and data shown in the figure.

8.2.1　*Data assimilation*

The data assimilation component employs data assimilation algorithms to assimilate real-time wildfire measurement data. Data assimilation can serve two purposes for supporting simulation-based prediction/analysis of wildfire spread:

- **Dynamic state estimation:** A simulation of wildfire spread needs to start from an initial state of a fire. The most important state information of a wildfire is the fire front locations, which dynamically change over time. The fire front locations are spatial state variables pertaining to the perimeter of the fire. There is a need to estimate the fire front locations because measurement data are noisy and provide only partial observations of a large wildfire area.

- **Online model calibration:** Wildfires often show different characteristics in different conditions, e.g., in mild or extreme weather conditions. Online model calibration makes it possible to adjust the model parameters to better characterize a specific fire. Furthermore, some parameters may dynamically change their values during the course of a fire spread. For example, the live fuel moisture content may dynamically change because a burning fire has the effect of curing nearby vegetations. For these parameters, it is best to treat them as dynamic parameters and estimate their values based on real-time measurement data collected from the real fire.

Due to the highly nonlinear non-Gaussian behavior of wildfire spread and the discrete event simulation model used for fire spread simulation, the data assimilation presented in this chapter will be based on the particle filter algorithm. More details will be provided later.

8.2.2 *Wildfire spread simulation model*

The wildfire spread simulation model simulates the spatiotemporal behavior of wildfire spread. It is used by both the data assimilation component and the simulation-based prediction/analysis component. The former uses the model as the state transition model for data assimilation. The latter uses the model to predict future wildfire spread.

Various wildfire spread simulation models have been developed in the literature. These models can be broadly categorized into physical (or theoretical) models and empirical or semi-empirical models. A survey of these models can be found in the works of Sullivan (2009a, 2009b). The physical models are based on fundamental physics involved in the combustion of biomass fuels. The empirical or semi-empirical models use statistical correlation between variables known to influence fire spread and field observations of rates of spread. In this work, we use the DEVS-FIRE model (described later), which is a semi-empirical model for surface fire spread simulation.

Despite many years of research, accurate simulation of wildfire spread remains a challenging task. Many models are fit in conditions that are too mild to be useful for extreme fire weathers

(McCaw *et al.*, 2008; Cruz and Alexander, 2013). The coupling effect between a spreading fire and its atmosphere is another factor that is often not adequately modeled. Furthermore, there is a lack of precise fuel and weather data, which are needed for wildfire spread simulation. For example, even though weather data may be obtained from nearby weather stations, the dynamic interactions between fire and weather often result in its own microscale "fire-weather" that is impractical to be measured accurately. Similarly, the characteristics of fuels vary from season to season, and some parameters (e.g., moisture content) can vary from day to day. This makes the fuel data that are measured at fixed times of a year inaccurate. All these factors limit the fidelity of wildfire spread simulations. Incorporating real-time measurement data is essential in order to support simulation-based prediction/analysis of wildfire spread.

8.2.3   *Real-time wildfire measurement data*

Real-time wildfire measurement data capture information about the real-time state of a spreading fire. To support operational simulation-based prediction, these data need to be collected in real time and updated frequently, e.g., multiple updates per hour.

Real-time measurement data may be collected using different methods, such as satellites, ground sensors, and manned aircrafts. Each of these data collection methods have limitations. For example, satellite data can cover large areas. However, they generally have infrequent updates (e.g., one update per day) and coarse spatial resolutions. As a result, satellite data have limited usage to support real-time prediction/analysis of wildfire spread. Ground fire sensors and manned aircraft can provide higher-resolution data with more frequent updates. However, ground fire sensors are difficult to deploy on demand and can be damaged by fires. Manned aircraft have limitations in terms of mission duration, mission safety, and cost. A relatively new technology that holds great potential for real-time wildfire data collection is *unmanned aircraft system* (UAS). UAS has the ability to fly on demand, operate in situations that are dangerous or too costly for manned aircraft, and continuously monitor an area to collect high-resolution data.

Regardless of the data collection methods, wildfire measurement data are often partial and noisy. The partial observations are due to the fact that a wildfire may spread across a large area that is too large to be fully covered by limited sensing resources. For example, a manned aircraft or UAS may fly around a fire area to collect data. At any time, the collected data cover only a subregion of the fire area. The noisy observations are due to the challenging wildfire environment (e.g., heavy smoke can block a UAS' view of the fire front) and the accuracy/precision constraints of sensors (e.g., limited resolution of thermal images). Advanced data assimilation methods are needed in order to effectively assimilate the partial and noisy measurement data.

8.2.4 *External inputs modeling and forecasting*

Wildfire spread prediction needs to take into account the future external inputs that can influence fire spread behavior. The two most important types of external inputs that impact a future fire spread are:

- **Forecasted weather condition:** Weather has a dynamic impact on wildfire spread. Although coupled atmosphere–fire simulation models have been developed, the majority of wildfire spread simulation models treat weather as an external input, whose data may be obtained from a nearby weather station. When using these models to predict/analyze a future fire spread, it is necessary to obtain forecasted weather conditions as the input for the simulation runs. The forecasted weather conditions can be obtained from a weather forecast service or from a separate weather model.
- **Planned fire suppression efforts:** Efforts of fire suppression can have major impacts on how a fire spreads. For a large-scale wildfire, it is common for firefighters to construct firelines (or firebreaks) at designated regions of the fire area while the fire is spreading. A simulation-based prediction/analysis thus needs to include not only the firelines that have already been constructed up to the current time but also the planned fire suppression efforts that will result in new firelines in the future. For example, if a new group of firefighters will be dispatched to a specific region to construct firelines, the simulation-based prediction/analysis should take that into consideration when predicting/analyzing the future fire spread. Fire suppression is a dynamic process by itself (e.g.,

the firelines are constructed based on the production rates of fire-fighting resources) and have a dynamic impact on fire spread. This means the simulation needs to have the capability of modeling and simulating the planned fire suppression efforts. Our previous work (Hu and Sun, 2007) developed agent-based modeling and simulation for wildfire suppression using realistic tactics.

8.2.5   *Simulation-based prediction/analysis*

The simulation-based prediction/analysis uses the simulation model to predict/analyze future fire spread. To achieve accurate prediction/analysis, the simulation runs are initialized from the fire front locations estimated from the *dynamic state estimation* and use the calibrated model parameters resulting from the *online model calibration*. Furthermore, the simulation runs need to take into consideration the forecasted weather conditions and planned fire suppression efforts computed from the *external inputs modeling and forecasting*.

The time window (i.e., how far ahead to predict/analyze the future) for a simulation-based prediction/analysis depends on its goals. For example, when the goal is to provide situation awareness for firefighters on the ground, predicting/analyzing fire spread in the next 30 minutes or 1 hour is appropriate for notifying firefighters in advance if they will be in danger. When the goal is to help decision-making in evacuating nearby neighborhoods, the time window may be set to 24 hours or more. In general, the further ahead into the future, the less accurate the simulation-based prediction/analysis would be.

A practical issue of simulation-based prediction/analysis is the requirement of finishing the simulations in real time. For wildfire simulation models that have high computation costs, parallel and distributed computing may be needed in order to satisfy this requirement.

The above discussions provide a general view of applying DDDS to wildfire spread simulation. The rest of this chapter focuses on the data assimilation works that have been carried out by the author's research group. These works are based on the DEVS-FIRE simulation model and assume that the measurement data are temperature data collected from ground temperature sensors.

8.3 The DEVS-FIRE Simulation Model

DEVS-FIRE (Ntaimo *et al.*, 2008, Hu *et al.*, 2012) is an integrated surface fire spread and suppression simulation model built on the discrete event system specification (DEVS) (Zeigler, 1976; Zeigler *et al.*, 2000). In DEVS-FIRE, the forest is modeled as a two-dimension cell space of rectangular cells whose dimensions depend on the resolution of the GIS fuel and topography data. Each cell performs its own local computation of the rate of spread and direction based on its fuel, topography, and prevailing weather conditions. Fire spread is modeled as a propagation process involving burning cells that ignite their unburned neighboring cells. Firefighting resources (e.g., dozers, fire crews) are modeled as firefighting agents, and fire suppression is modeled as a process of constructing firelines to suppress or contain a burning fire. Firefighting agents execute fire suppression plans by interacting with the forest cells through message passing.

Figure 8.3 shows the structure of the DEVS-FIRE model, where the components in gray color are implemented as DEVS atomic or coupled models. At the center of DEVS-FIRE is the *Forest Cell Space* coupled model. This model comprises a grid of *forest cells*, with the fuel, topography, and weather conditions assumed to be uniform within the cell. Each cell is a DEVS atomic model; it transitions through different states (e.g., *unburned, burning, burned*) during the simulation. When ignited, a cell uses Rothermel's fire behavior model

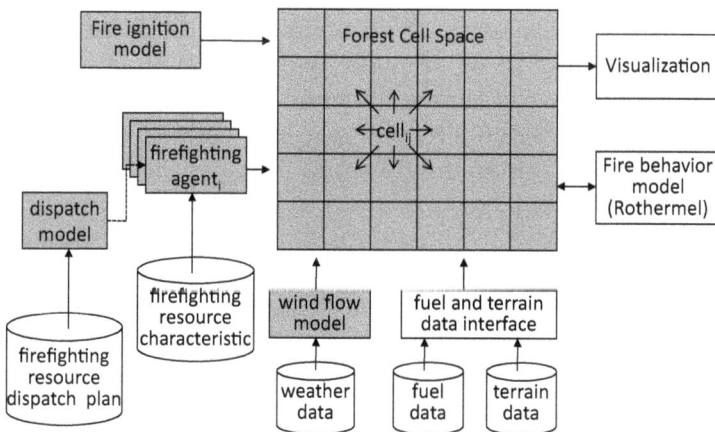

Figure 8.3.   Structure of the DEVS-FIRE model.

(Rothermel, 1972) to compute a one-dimensional fire spread (speed and direction), which is then decomposed into two dimensions based on an elliptical fire shape.

Each cell has input and output ports through which couplings are made for exchanging messages. A cell is coupled to its eight adjacent neighboring cells, as illustrated by $cell_{ij}$ in the figure. Through these couplings, fire spread across the cell space is enabled via message exchange between neighboring cells. Topography data (including slope data and aspect data) and fuel data are read through a topography and fuel data interface. This allows each forest cell to be initialized with the correct fuel and topography data. A cell receives weather data through a *wind flow* atomic model, which is coupled to all the cells for updating weather information. The *Fire ignition* atomic model is responsible for igniting an initial set of cells to start the simulation. The *visualization* component displays the simulation results. It changes a cell's display color whenever the cell's state changes.

The *firefighting agent* models and a *dispatch* model are used to support fire suppression simulation. A firefighting agent is a DEVS atomic model modeling a firefighting resource. Its characteristics, such as fireline production rate and start time, are based on the actual resource it represents. Multiple firefighting agents can be added to a fire suppression simulation to model multiple groups of firefighting resources. DEVS-FIRE supports three types of firefighting agents that model different fire suppression tactics, including direct attack, parallel attack, and indirect attack (Fried and Fried, 1996). Details of the firefighting agents can be found in the works of Hu and Sun (2007) and Hu and Ntaimo (2009). The dispatch model is responsible for creating firefighting agents and adding them to the simulation. It reads from a firefighting resource dispatch plan and constructs firefighting agents to execute the plan.

DEVS-FIRE uses a modular design in which the cell space model simulates the dynamics of wildfire spread and the firefighting agents simulate the fire suppression effort. This design separates the concerns of fire spread and fire suppression and makes it easy to evolve each one. A fire spread simulation can be run without involving the

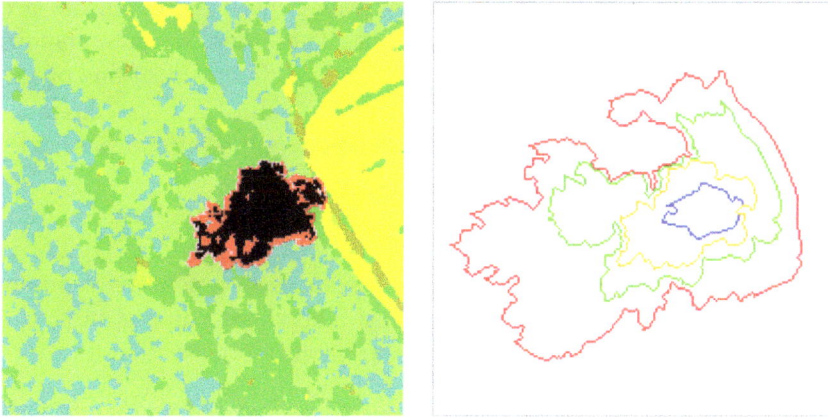

Figure 8.4. Fire spread simulation using the DEVS-FIRE model: The left-hand figure shows a snapshot of a fire spread simulation. Black cells are burned cells, red cells are burning cells, and all other colors represent different fuel types of the area. For better visual effect, the grid lines of the cell space are not displayed in the figure. The right-hand figure shows how the fire grows over time by drawing the fire perimeters (in different colors) at different time instants of the fire spread simulation.

fire suppression model components. This chapter focuses on data assimilation for fire spread simulation and thus uses only the fire spread model of DEVS-FIRE.

DEVS-FIRE is a deterministic simulation model. When starting from the same initial state and receiving the same input trajectory, different simulations would produce the same simulation results, i.e., generate the same state trajectories and output trajectories.

Figure 8.4 shows an example of fire spread simulation using the DEVS-FIRE model.

8.4 Temperature Measurement Data

The measurement data are temperature data collected from ground temperature sensors. The temperature sensors are deployed in the wildfire area based on a deployment schema. Examples of deployment schema include regular deployment, e.g., one sensor every certain distance away; random deployment, where sensors are randomly distributed; and fire-directed deployment, where more sensors are deployed around the active fire regions. These deployment schemas (and the total number of sensors) result in different locations of the sensors. It is assumed the number of sensors and their specific locations are known. Furthermore, we consider only stationary sensors that do not change locations over time.

The measurement vector includes the temperature data from all the sensors. The sensors report their temperature readings regularly. The readings are influenced by the distances between the temperature sensors and the active fire front: The closer the distance, the higher the temperature (see the measurement model section for more details). For sensors that are far away from the fire front, their temperature readings are close to the ambient temperature. Since sensors are spatially distributed and the fire front evolves over time, the temperature readings from different sensors change over time as the fire spreads.

Figure 8.5 illustrates the measurement data for a given fire state at a specific time. Three sensor deployment schemas are considered and illustrated. Figure 8.5(a) illustrates the overall temperature map for a wildfire area. Due to the influence of the fire, different locations have different temperatures, displayed in colors varying continuously from black to red, and then to yellow and white. The active fire front locations have the highest temperatures, while locations (including the ones that are already burned out) far away from the fire front have lower temperatures. Figures 8.5(b)–(d) show the temperature measurement data collected from sensors that are deployed based on three different deployment schemas, in which the sensors' temperature values are denoted by the shades of red (the darker the red, the higher the temperature). In Figure 8.5(b), 100 sensors are regularly deployed. Figures 8.5(c) and (d) use the same number of sensors. However, in these two cases, sensors are randomly deployed (referred to as *random deployment 1* and *random deployment 2*, respectively).

Note that while the measurement data carry information about the active fire front, the precise fire front locations cannot be directly derived from the data. The goal of data assimilation is to assimilate the temperature data to estimate the dynamically changing fire front locations.

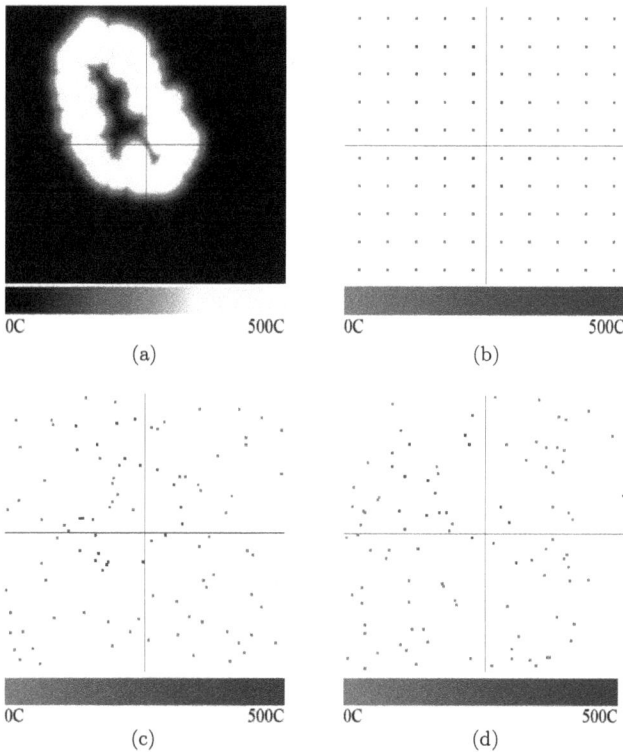

Figure 8.5. Illustration of temperature measurement data: (a) overall temperature map; (b) measurement data with 100 sensors regularly deployed; (c) measurement data with 100 sensors randomly deployed (named as *random deployment schema 1*); (d) measurement data with 100 sensors randomly deployed (named as *random deployment schema 2*).

8.5 Data Assimilation for Dynamic State Estimation

Due to the nonlinear non-Gaussian behavior of wildfire spread as well as the discrete event DEVS-FIRE simulation model, we use particle filters to carry out the data assimilation to estimate fire front locations.

8.5.1 *Problem formulation*

The DEVS-FIRE model uses cells to represent a fire area. Based on the states of individual cells, the fire front locations can be extracted as the locations of the cells that are in the *burning* state. In this work, the state of a wildfire, referred to as the *fire state*, is composed of the states of all the cells in the cell space. Formally, we define the fire state as an n_c-dimensional vector:

$$fire = [x_1, \ldots, x_{n_c}]^T, \tag{8.1}$$

where $x_i \in \{unburned, burning, burned\}$ is the state for the ith cell $(i = 1, \ldots, n_c)$ and n_c is the total number of cells in the whole cell space.

Accordingly, the DEVS-FIRE simulation model can be described as a function that evolves the fire state over time, i.e.,

$$fire_{t+\Delta t} = DEVSFIRE(fire_t, \theta, u_{t+\Delta t}, \Delta t), \tag{8.2}$$

where $DEVSFIRE()$ stands for the DEVS-FIRE simulation model, $fire_t$ and $fire_{t+\Delta t}$ are the fire states at times t and $t + \Delta t$, respectively, θ is a vector containing the model parameters, $u_{t+\Delta t}$ is the external input between times t and $t + \Delta t$, and Δt is the forward simulation time. In this work, the external input is the weather input that includes wind speed and wind direction. Note that θ does not have a subscript t because we assume static parameters in this work.

The fire state defined in Equation (8.1) treats each cell's state as a discrete categorical state, i.e., $x_i \in \{unburned, burning, burned\}$. We note that a cell in the DEVS-FIRE model actually includes other state variables, such as the elapsed time in the burning state, which are not included in the above state definition. In other words, the cell state $x_i \in \{unburned, burning, burned\}$ is a scaled-down version of the actual state of a cell in the DEVS-FIRE model. This choice of using a scaled-down state for each cell is to reduce the size of the overall state vector that needs to be estimated in data assimilation. For the data assimilation problem considered in this work, the fire state defined above captures the most important information about the fire spread condition. The loss of information due to not including the other state variables is minimal and thus can be handled through process noise of the state transition model (described later). We note that this treatment is acceptable only when there is minimal loss of information due to not including the other state variables.

Let t and $t + \Delta t$ be the time instants at data assimilation steps $k-1$ and k. Equation (8.2) can be rewritten in a way consistent with the notations used in Chapter 6:

$$fire_k = DEVSFIRE(fire_{k-1}, u_k), \qquad (8.3)$$

where $fire_{k-1}$ and $fire_k$ are the fire states at data assimilation steps $k - 1$ and k and u_k is the external input at step k. Equation (8.3) omits θ because it is static and can be treated as part of the DEVS-FIRE model. It also omits Δt, which is the time interval between step $k - 1$ and k.

The measurement data include the temperature data from all the temperature sensors deployed in the fire area. Let n_s be the total number of sensors. The measurement vector y is an n_s-dimensional vector containing temperature values from all the sensors:

$$y = [y_1, \ldots, y_{n_s}]^T, \qquad (8.4)$$

where $y_i \in R$ is a real number representing the temperature reading from the ith sensor $(i = 1, \ldots, n_s)$. The measurement vector at data assimilation step k is denoted by y_k.

Based on the above definitions, we formulate the nonlinear state-space model for the wildfire data assimilation problem as follows:

$$fire_k = DEVSFIRE(fire_{k-1}, u_k) + \gamma_k, \qquad (8.5)$$

$$y_k = MF(fire_k) + \varepsilon_k. \qquad (8.6)$$

Equation (8.5) is the state transition model, where $DEVSFIRE$ $(fire_k, u_k)$ is the DEVS-FIRE simulation model described above, and γ_k is the process noise that introduces noises (or uncertainties) in the state evolution. Equation (8.6) is the measurement model, where y_k is the measurement vector, $MF(fire_k)$ is the measurement function mapping fire state to measurement, and ε_k is the measurement noise. Details of the measurement function will be described later.

8.5.2 *Sampling using the state transition model*

The state transition model is used by the particle filter algorithm to generate new samples of fire state. As shown in Equation (8.5), the state transition model includes the DEVS-FIRE simulation model and a noise factor γ_k. The noise factor γ_k models the process noise involved in the state transition. Since this work focuses on estimating the fire front locations, we define γ_k as a graph noise over a fire shape. A method of adding graph noise for a given fire shape is developed. In this method, a fire shape perimeter is first divided into multiple segments, each of which consists of an equal number of burning cells. For each segment, we then introduce a noise level that defines the amount of change (in number of cells) that shifts cells of the segment inside or outside along the direction from the ignition point to the cells. Different segments may have different noise levels, but all cells of the same segment share the same noise level. Each cell of a segment is then shifted to a new position according to the noise level of the segment. The shifted cells are reconnected to form a new fire front called the *noised fire front*.

To illustrate the effect of the graph noise, Figure 8.6 shows the results of four different runs of the graph noise method for a given fire state. In each run, we compare the original fire front with the noised fire front.

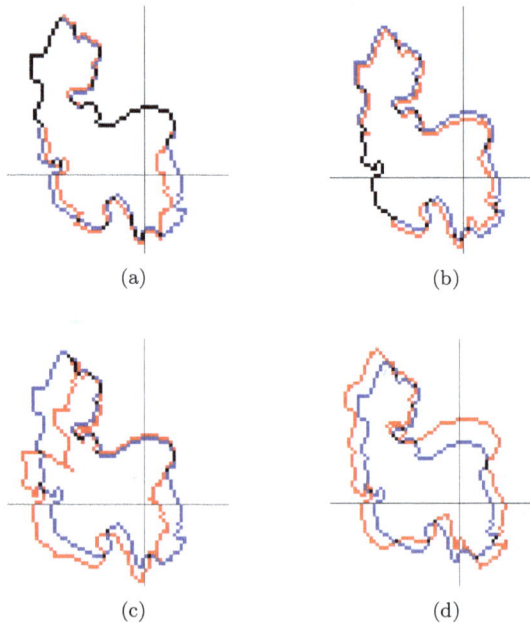

(a) (b)

(c) (d)

Figure 8.6.   Noised fire fronts generated from four different runs of the graph noise method for a given fire state. In each case, the original fire front is shown in blue and black (the black cells are on both the original fire front and the noised fire front); the noised fire front is shown in red and black.

Based on the DEVS-FIRE model and the graph noise method described above, the sampling procedure includes first running a simulation using the DEVS-FIRE model and then adding graph noise to the simulated fire shape. Specifically, for each particle, we create a DEVS-FIRE simulation starting from the fire front defined by that particle and run the simulation for the time that equals the data assimilation step interval. Let $\widetilde{fire}_k$ denote the result of $DEVSFIRE(fire_{k-1}, u_k)$. To add graph noise to the simulated fire shape, we first extract the fire front perimeter from $\widetilde{fire}_k$ and then add graph noise to this fire front. We then set all the cells on the noised fire front to the *burning* state, all the cells inside the noised fire front to the *burned* state, and all the cells outside the noised fire front to the *unburned* state. The resulting states of all the cells become the new sample, denoted by $\overline{fire}_k$, for the particle. The specific sampling algorithm, including details of the graph noise method, can be found in the work of Xue *et al.* (2012).

The graph noise of the sampling procedure introduces uncertainty into the state transition. This is necessary because the DEVS-FIRE model is a deterministic model, which means different particles starting from the same state will lead to the same sampling result if no uncertainty is added. The graph noise also serves as the process noise of the state transition. The process noise accounts for the errors that exist in the DEVS-FIRE model to make data assimilation generate more robust estimation results.

8.5.3 *Measurement model*

The measurement model (Equation (8.6)) maps the fire state to the measurement vector. It includes the measurement function $MF(fire_k)$ and the measurement noise ε_k.

The measurement function $MF(fire_k)$ computes the temperature data of deployed sensors for a given fire state. For a specific sensor and a given fire state, the function computing the sensor's temperature is defined by

$$T = T_c e^{-\frac{1}{2}\left(\frac{d}{\sigma}\right)^2} + T_a, \tag{8.7}$$

where T is the temperature of the sensor, T_c (°C) refers to the temperature rise above the ambient temperature of the closest burning cell on the fire front, T_a denotes the ambient temperature (°C), d denotes the distance from the sensor to the closest burning cell, and σ is a constant and is set to 50 m in our work. To compute T_c for a burning cell, Equation (8.8) is used (Van Wagner, 1973, 1975):

$$T_c = 3.9 \ FI^{\frac{2}{3}}/h. \tag{8.8}$$

In Equation (8.8), FI is the fireline intensity of the burning cell (kW/m) and h is the height above ground (m). In the DEVS-FIRE simulation, the fireline intensity (FI) of a burning cell is computed from Rothermel's model at runtime. For ground temperature sensors, the height h would be their installation heights.

Figure 8.7 illustrates how the measurement function $MF(fire_k)$ works in computing the temperature data of ground temperature sensors. It displays a simplified fire state, where eight cells are burning (displayed in red) in a 9×9 cell space, with each cell's size being 15 m. The temperature sensors are regularly deployed in the cell space with one temperature sensor every three cells (in both horizontal and vertical directions). The cells where sensors are deployed are displayed in gray. To illustrate how the temperature data are calculated, in the following, we assume all burning cells' T_c is 376°C and the ambient temperature is 27°C. From Equation (8.7), we have $T = 376e^{-d^2/2\sigma^2} + 27$ ($\sigma = 50$), where d is the distance from a sensor to its closest burning cell on the fire front. Based on this formula, we can obtain the temperature data of all the sensors, which are $\{149, 267, 267, 267, 386, 386, 267, 386, 371\}$ (°C) indexed from left to right and from top to bottom.

The measurement noise ε_k adds noises to the temperature data computed from $MF(fire_k)$. In this work, the measurement noise is modeled as a zero-mean Gaussian noise for each sensor, i.e., $\varepsilon_{j,k} \sim N(0, \sigma_\varepsilon^2)$, where j is the index for the jth sensor and $\sigma_\varepsilon = 20.0$ is a constant for all sensors.

The measurement model defined above means that the measurement data are *scalar Gaussian measurements*, as defined in Section 6.6.1. This work assumes sensor data are independent of each other. Using Equation (6.36) defined in Chapter 6, the likelihood

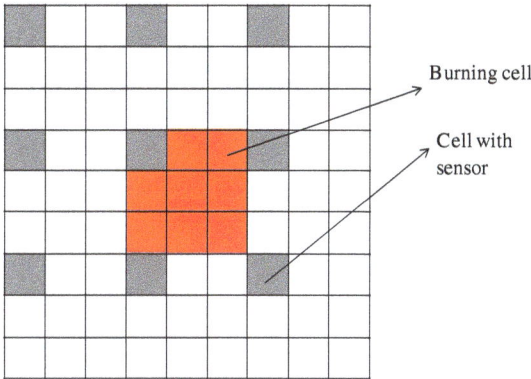

Figure 8.7. An example of the measurement function $MF(fire_k)$.

probability of observing measurement data y_k^{md} for a given state $fire_k$ is defined as

$$p(y_k = y_k^{md} | fire_k) = \prod_{j=1}^{n_s} p(y_{j,k} = y_{j,k}^{md} | fire_k), \qquad (8.9)$$

$$p(y_{j,k} = y_{j,k}^{md} | fire_k) = \frac{1}{\sigma_\varepsilon \sqrt{2\pi}} e^{-\frac{1}{2}\left(\frac{MF_j(fire_k) - y_{j,k}^{md}}{\sigma_\varepsilon}\right)^2}, \qquad (8.10)$$

where n_s is the total number of sensors, $y_{j,k}^{md}$ is the real measurement data from the jth sensor, and $MF_j(fire_k)$ is the predicted measurement for the jth sensor using the measurement function $MF(fire_k)$ described above. The likelihood probability is used to compute the importance weights of particles.

8.5.4 *The particle-filter-based data assimilation algorithm*

Particle filters are used to assimilate the temperature data into the DEVS-FIRE simulation model. Algorithm 8.1 shows the particle-filter-based data assimilation for wildfire spread simulation, where N is the total number of particles and y_k^{md} is the temperature measurement data at step k.

Algorithm 8.1. Particle- Filter-based Data Assimilation for Wildfire Spread Simulation

1. Initialization, $k = 0$:

   - for $i = 1, \ldots, N$, sample $fire_0^{(i)}$ based on information of the initial fire state.
   - Advance the time step by setting $k = 1$.

2. Sampling:

   - For each fire state in $\{fire_{k-1}^{(i)}\}_{i=1}^N$, draw a sample (denoted by $\overline{fire}_k^{(i)}$) using the sampling procedure described in Section 8.5.2, which includes running a DEVS-FIRE simulation and then adding graph noise.

3. Importance weight computation:

   - For each fire state in $\{\overline{fire}_k^{(i)}\}_{i=1}^N$, compute its importance weight: $\bar{w}_k^{(i)} = p(y_k = y_k^{md}|\overline{fire}_k^{(i)})$ using Equation (8.9) based on the temperature measurement data y_k^{md}.
   - Calculate the normalized weights: $w_k^{(i)} = \dfrac{\bar{w}_k^{(i)}}{\sum_{j=1}^N \bar{w}_k^{(j)}}$.

4. Resampling:

   - Resample N particles $\{fire_k^{(i)}\}_{i=1}^N$ according to $\{w_k^{(i)}\}_{i=1}^N$ using the resampling algorithm (Algorithm 6.3).

5. Advance the time step and start a new cycle:

   - Set $k \leftarrow k + 1$, and go to step 2.

This algorithm is a specific implementation of the bootstrap filter algorithm (Algorithm 6.4). The *Importance sampling step* of Algorithm 6.4 is broken into two steps (*Sampling* and *Importance weight computation*) to explicitly show the sampling that uses the simulation model and the weight computation that uses the measurement data.

The algorithm starts by initializing N particles representing the initial belief of the fire states. In this work, we start the data assimilation from the time when the fire is ignited and initialize all the particles using the same ignition cell. Then, the algorithm goes through the data assimilation cycles, each of which includes the sampling, importance weight computation, and resampling steps. In the sampling step, all the particles go through the sampling procedure to generate a new noised fire front. In the importance weight computation step, the weights of all the samples are computed based on the temperature measurement data; these weights are then normalized. In the resampling step, a new set of particles is selected based on their normalized weights. These particles represent the belief of the new estimated fire state after assimilating the data of that step. They are also used as the input for the next step of data assimilation.

8.5.5 *Data assimilation results*

We use the identical twin experiment described in Section 7.4 to evaluate the data assimilation for the wildfire spread simulation. In the following description, we use three terms: *real fire*, *filtered fire*, and *simulated fire*, to help present the experiment results:

- A real fire is the simulated fire spread from which the real measurement data are obtained. The real fire also allows the recorded "true" state to be compared with the data assimilation results.
- A simulated fire is the simulation result based on some "erroneous" data ("erroneous" in the sense that the data are different from those used in the real fire), e.g., imprecise weather data. This is to represent the fact that wildfire simulations usually rely on imperfect data as compared to real wildfires.
- A filtered fire is the data assimilation result based on the same "erroneous" data as in the simulated fire. In our experiment, each data assimilation step chooses the particle with the largest importance weight before resampling as the filtered fire. Our goal is to show that a filtered fire gives more accurate results by assimilating measurement data from the real fire, even though it still uses the "erroneous" data as in the simulated fire.

The differences between a real fire and a simulated fire are due to the imprecise data used in the simulation. In this experiment, we choose to use imprecise wind conditions (wind speed and wind direction) as the "erroneous" data. Table 8.1 shows the configurations of four experiment cases. The wind speed and direction used in the real

Table 8.1. Wind data used in the experiment.

	"Erroneous" data		Data used in the "real fire"	
Case	Speed (mph)	Direction (°)	Speed (mph)	Direction (°)
1	6 ± 2	No error	8 ± 2	180 ± 20
2	10 ± 2			
3	No error	160 ± 30		
4		200 ± 30		

fire are 8 mph and 180°, respectively, with random variances added every 10 min. The variances for the wind speeds are in the range of −2 to 2 mph (denoted as 8 ± 2 in the table), and the variances for the wind direction are in the range of −20 to 20 degrees (denoted as 180±20 in the table). The first two experiment cases introduce errors to the wind speeds only, and the next two experiment cases introduce errors to the wind directions only. The corresponding "erroneous" data are shown in the table.

All simulations use the GIS data and fuel data acquired from Huntsville, Texas, USA, during the leaf-off season in March 2004. The cell space dimension is 200 × 200 and the cell size is 15 m. The ignition point is set to the center point of the cell space for all of the simulations. The measurement data (ground temperature sensor data) from the real fire are collected every 20 min. Sensors are assumed to be regularly deployed in the cell space, with one sensor in every 10 cells. The particle filter algorithm uses 50 particles.

Figure 8.8 shows the data assimilation results after eight steps (20 min in each step) of the data assimilation for the four experiment cases. One can see that in all four cases, the filtered fires match the real fire better than the simulated fires do. Taking case 1 as an example, due to the erroneous smaller wind speeds, the simulated fire is significantly smaller than the real fire at the head of the fire (the top part of the fire shape). Using data assimilation, the filtered fire overcomes this problem and matches the real fire with a smaller difference. Similar results can be observed in all other cases. These results indicate that the data assimilation can effectively estimate the true fire state and thus improve simulation results.

More in-depth analysis and quantitative results of the data assimilation can be found in the work of Xue *et al.* (2012). It is worth

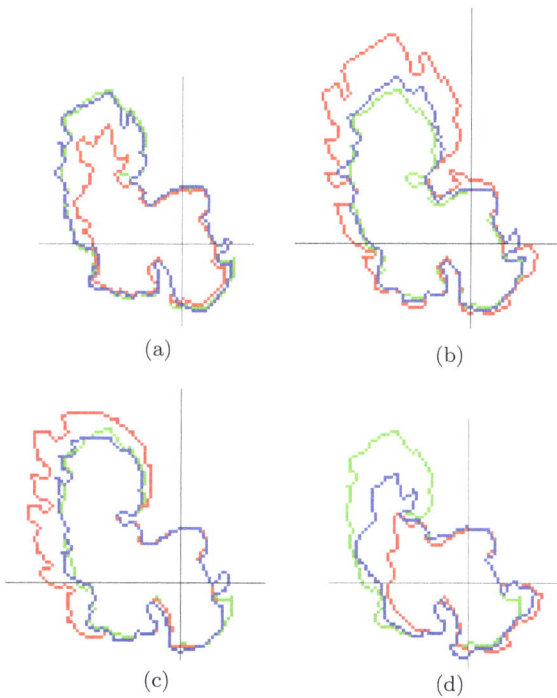

Figure 8.8. Data assimilation results by comparing fire fronts from the real fire (green), the simulated fire (red), and the filtered fire (blue): (a) case 1; (b) case 2; (c) case 3; (d) case 4.

mentioning that the data assimilation results depend on many parameters, such as the number of particles, number of sensors, and frequency and quality of sensor data. An analysis and quantification of the particle-filter-based data assimilation for wildfire spread simulation can be found in the work of Gu and Hu (2010).

8.6 Spatial Partition-Based Particle Filter

8.6.1 *Motivation*

The data assimilation described in the previous section follows closely the bootstrap filter algorithm. In particular, it treats the fire state as a whole and computes the importance weight of each particle using all the sensors.

A unique property of many spatiotemporal systems, including the wildfire spread application, is *spatial locality*. Spatial locality means that the measurement data and system states are spatially dependent, and their immediate influences are constrained to their local areas. In the wildfire application, fire spreads to their neighboring cells, and the measurement data (e.g., ground temperature sensor data) only reflect the fire states in their nearby areas. Since both the fire state and measurement data are spatially dependent and have finite correlation lengths, a particle filter algorithm that considers the fire state as a whole and calculates the particles' weights using all the sensors can result in situations where the particles' importance weights do not accurately reflect some local states.

Figure 8.9 illustrates this problem. In the figure, Particle 1's (the left-hand figure) overall fire front (dotted) deviates greatly from the true fire front (continuous). As a result, this particle would be assigned a low importance weight based on the weight computation that uses all the sensor data. Nevertheless, one can see that a portion of the particle's fire front located within the orange circle matches well with the true fire front. However, this portion of the fire state (i.e., the local state) has a high chance of being eliminated along with the overall particle in the resampling step due to the small weight of the particle. On the other hand, Particle 2's (the right-hand

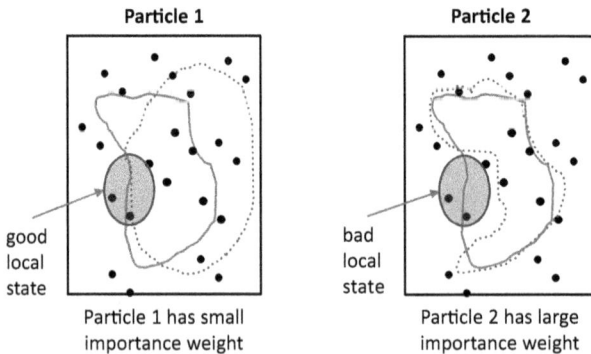

Figure 8.9. Illustration of the problem that a particle's overall weight does not accurately reflect some local states. The continuous line is the true fire front; the dotted line is the fire front represented by a particle; the dot points are location of ground temperature sensors that are randomly distributed.

figure) fire front largely matches the true fire front and thus would be assigned a large weight. This particle is likely to be maintained and duplicated multiple times through resampling, even though its local state within the orange circle is worse than that of Particle 1. The above problem can lead to a result where none of the particles after resampling has a good representation of the local state within the orange circle.

One way to address the above problem is to use a large number of particles, which would increase the chance of having good representations of all the local states. However, the large number of particles lead to a higher computational cost.

An alternative way is to exploit the spatial locality of state and measurement data. The spatial locality makes it possible to employ a divide-and-conquer strategy to partition the overall space into smaller sub-areas and then design efficient sampling and resampling procedures for each sub-area. This leads to the *spatial partition-based particle filter* described in the following.

8.6.2 *The spatial partition-based particle filter*

The basic idea of the spatial partition-based particle filter is to split the state and measurement data into sub-components according to sub-areas of the overall space and perform weight computation and resampling based on the sub-components. Similar to the standard particle filter, the spatial partition-based particle filter includes sampling, weight computation, and resampling in each cycle. It also includes an extra step of *state partition* that happens after the sampling step. Sampling is still based on the full fire state because the DEVS-FIRE model needs the full state to simulate fire spread. However, unlike the standard particle filter, weight computation is based on sub-states and uses only the measurement data from the sensors that have observation coverage over the sub-states. Resampling also happens at the sub-state level to reconstruct a full state based on the importance weights of sub-states.

Let the whole fire area be divided into m sub-areas, i.e., $r = r_1 \cup r_2 \cup \cdots r_i \cdots \cup r_m$, where r is the whole space and r_i for $i = 1 \ldots, m$ are sub-areas. According to the sub-areas, we can partition the full state x into sub-states and the measurement data y into sub-measurements. Figure 8.10 shows an example of a grid-based

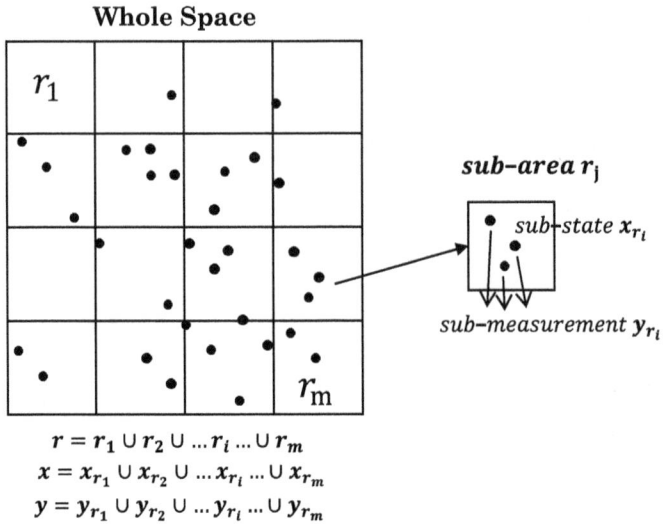

Figure 8.10.   A grid-based spatial partition and the generated sub-states and sub-measurements. The black dots are locations of the ground temperature sensors.

spatial partition. In the figure, r_i is a sub-area, x_{r_i} is the sub-state corresponding to r_i, and y_{r_i} denotes the sub-measurement related to r_i.

Based on the spatial partition described above, the spatial partition-based particle filter includes four steps in each data assimilation cycle. The four steps are illustrated in Figure 8.11 and described as follows.

Step 1: Sampling: The sampling step is the same as that in the standard particle filter. It employs the sampling procedure described in Section 8.5.2 to generate a new full fire state for each particle. Because a sub-area's fire can spread to other sub-areas, sampling does not happen at the sub-state level — it uses the full fire state to run the fire spread simulation. A sampled state is denoted by $\bar{x}_t$ in Figure 8.11.

Step 2: State partitioning: The state partitioning step splits a particle's full fire state generated from the sampling step into sub-states (denoted by $\bar{x}_{r_i,t}$ in Figure 8.11) according to the spatial partitions of the fire area. A fire area may be partitioned into sub-areas using different methods, such as grid-based partitioning, geography-based partitioning, and sensor-based partitioning.

Sampling	State Partitioning	Weight Calculation	Resampling

$$\bar{x}_t \sim p(x_t|x_{t-1}) \longrightarrow \boxed{\begin{array}{c}\bar{x}_{r_1,t}\end{array}} \longrightarrow p(y_{r_1,t}|\bar{x}_{r_1,t}) \to / \longrightarrow * \to \boxed{\begin{array}{c}x_{r_1,t}\end{array}} \to x_t$$

$$\bar{x}_{r_2,t} \longrightarrow p(y_{r_2,t}|\bar{x}_{r_2,t}) \to / \longrightarrow * \to x_{r_2,t}$$

$$\vdots \qquad \vdots \qquad\qquad \vdots \qquad \vdots\ \vdots \qquad \vdots\ \vdots \qquad \vdots$$

$$\bar{x}_{r_i,t} \longrightarrow p(y_{r_i,t}|\bar{x}_{r_i,t}) \to / \longrightarrow * \to x_{r_i,t}$$

$$\vdots \qquad \vdots \qquad\qquad \vdots \qquad \vdots\ \vdots \qquad \vdots\ \vdots \qquad \vdots$$

$$\bar{x}_{r_m,t} \longrightarrow p(y_{r_m,t}|\bar{x}_{r_m,t}) \to / \longrightarrow * \to x_{r_m,t}$$

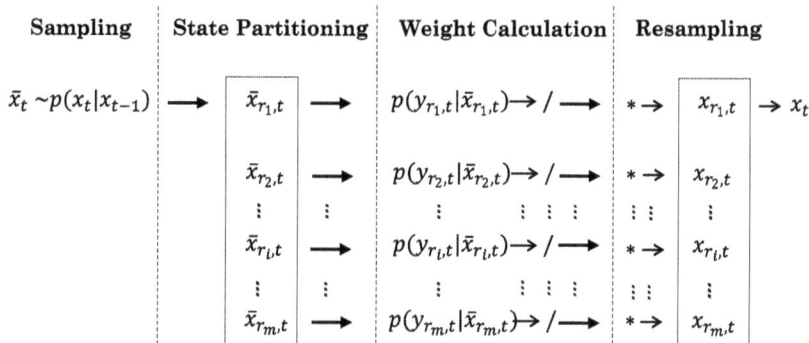

Figure 8.11. General flow of spatial partition-based particle filter. Weight normalization is denoted by /. Resampling is denoted by *.

The grid-based partitioning divides the area based on arbitrarily defined grids. The geography-based partitioning divides the area according to geographical features, such as roads, rivers, and land boundaries. The sensor-based partitioning divides the area based on spatial locations of sensors and related partition criteria (such as to make each sub-area have a balanced number of sensors). We developed in our previous work (Long and Hu, 2017) a two-level partitioning method to provide a spatial partition of a fire area that has a balanced number of sensors in each sub-area and fewer boundary sensors between sub-areas.

Step 3: Weight computation: The weight computation step differs significantly from that in the standard particle filter, as it computes the importance weight for each sub-state instead of the full fire state. The weight of each sub-state is computed as the likelihood $p(y_{r_i,t}|\bar{x}_{r_i,t})$, where $\bar{x}_{r_i,t}$ represents a sub-state that belongs to region r_i, and $y_{r_i,t}$ denotes the measurement data associated with region r_i. After the sub-states' weights from all the particles are computed, they are normalized for each sub-state (this is denoted by "/" in Figure 8.11). Note that the weight computation at the sub-state level needs to address the *data association* problem for boundary sensors, which are sensors that have observation coverages over multiple sub-areas. These boundary sensors can be influenced by multiple sub-states and thus lead to the issue of how to compute the likelihood probability of observing a boundary sensor's data due to a specific sub-state. The work of Long and Hu (2017) employs a strategy that computes the likelihood based on the assumption that all the related sub-states have equal probabilities of influencing a boundary sensor.

Step 4: Resampling (denoted by "*" in Figure 8.11): The goal of the resampling step is to construct a full fire state based on the sub-states and their importance weights. After the weight computation step, for each sub-area r_i, we have a set of normalized weights $\{w_{r_i,t}^{(1)}, \ldots, w_{r_i,t}^{(N)}\}$ corresponding to the set of sub-states $\{\bar{x}_{r_i,t}^{(1)}, \ldots, \bar{x}_{r_i,t}^{(N)}\}$ that are contributed from all the particles (N is the total number of particles). To resample a full fire state for a new particle, for each sub-area r_i, we resample a sub-state (denoted as $x_{r_i,t}$ in Figure 8.11) from $\{\bar{x}_{r_i,t}^{(1)}, \ldots, \bar{x}_{r_i,t}^{(N)}\}$ according to their weights and then combine the resampled sub-states from all the sub-areas to form a full fire state. The new set of particles from the resampling step represents the belief of the current state and is used as the input for the next step of data assimilation.

8.6.3 *Experiment results*

We use the identical twin experiment similar to the one described in Section 8.5.5 to evaluate the data assimilation using the spatial partition-based particle filter. The measurement data are collected from ground temperature sensors, as described in Section 8.5.3. The sensors are randomly distributed in the fire area. They report temperature data every 20 min.

Figure 8.12 shows the data assimilation results of the spatial partition-based particle filter. The experiments are based on 12 steps of data assimilation (each step is 20 min). The particle filter algorithms use 50 particles. Figure 8.12(a) shows the result from the standard particle filter (no spatial partition). Figures 8.12(b) and (c) show the results from the spatial partition-based particle filter when the space is partitioned into two and six sub-states, respectively. Figure 8.12(d) compares the mean square error (MSE) for the data assimilation results using different number of sub-states. The MSE is computed as the ratio of the cells that have different states between the filtered fire and the real fire.

From Figure 8.12, one can see that the simulated fire and the real fire differ significantly due to the erroneous weather data used by the simulation. Through data assimilation, the standard particle

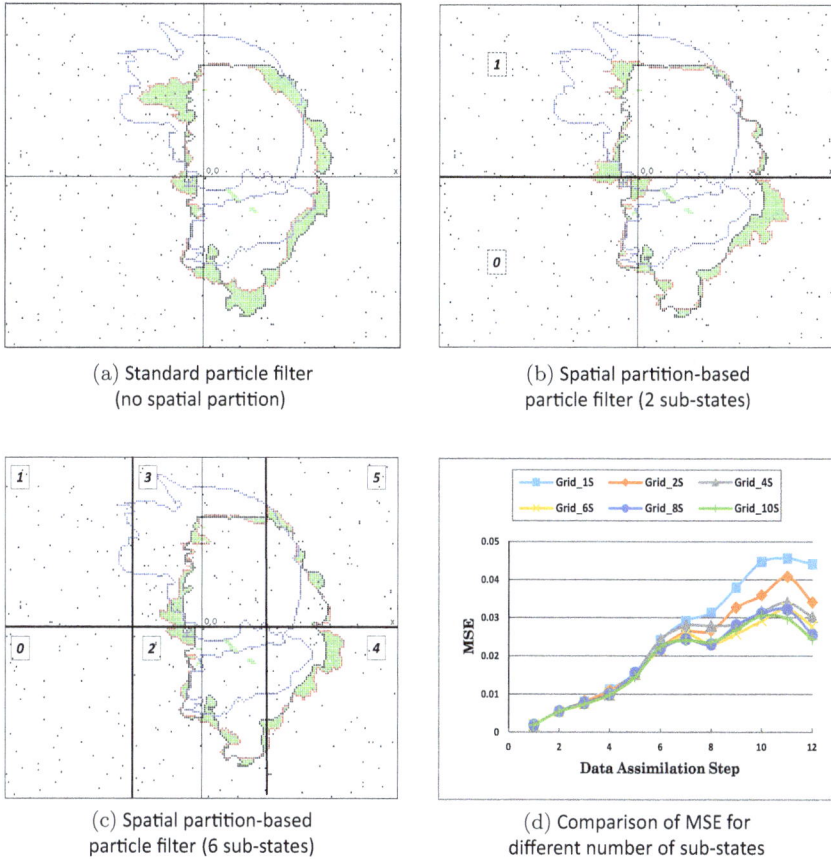

(a) Standard particle filter
(no spatial partition)

(b) Spatial partition-based
particle filter (2 sub-states)

(c) Spatial partition-based
particle filter (6 sub-states)

(d) Comparison of MSE for
different number of sub-states

Figure 8.12. Data assimilation results using the spatial partition-based particle filter: In (a)–(c), the blue line is the simulated fire, black line is the real fire front, red line is the filtered fire, green areas represent the difference between the filtered fire and the real fire; in (d), $Grid\_1S$ is the result from the standard particle filter, and $Grid\_2S$, $Grid\_4S$, ..., $Grid\_10S$ are the results from the spatial partition-based particle filter using 2, 4, ..., 10 sub-states, respectively.

filter is able to make the filtered fire roughly match the real fire. Nevertheless, there are still noticeable differences between the two fire fronts, as shown in Figure 8.12(a). Figure 8.12(b) shows that the spatial partition-based particle filter using two sub-states was able to significantly improve the degree of match between the two fire fronts (especially on the northern part of the fire). The result is further improved when using six sub-states, although the improvement is not as significant as that of the two sub-states. Figure 8.12(d) shows that at the 12th step of data assimilation, the standard particle filter has the highest MSE, which is significantly reduced when using two sub-states and then further reduced when using four sub-states. After the four sub-states, partitioning the space into more sub-states does not bring about significant improvements. These results show that the spatial partition-based particle filter is able to improve the data assimilation result. However, the benefit of adding more sub-states would decrease after a certain threshold is reached. More results and analysis can be found in the work of Long and Hu (2017).

8.7   Parallel/Distributed Particle Filtering

An important aspect of particle-filter-based data assimilation for wildfire spread simulation is the high computation cost that it involves. The high computation cost is due to two reasons. First, to estimate the high-dimensional state of a wildfire, the particle filter algorithm needs to use a large number of particles. During the sampling step, each particle runs a full-scale fire spread simulation to the next data assimilation time. Second, a wildfire spread simulation is computation intensive by itself. For example, when simulating a large wildfire, the DEVS-FIRE model would have a large number of cells that are in the burning state, each of which needs to compute the rate and direction of fire spread based on its local fuel, terrain, and weather data. The large number of particles and the computation-intensive simulation together result in the high computation cost for the particle-filter-based data assimilation.

To improve the performance of data assimilation, parallel/ distributed implementations of particle filters have been developed. Two issues have been identified for parallel/distributed particle filtering (Bolic *et al.*, 2005). First, the resampling of the standard particle filter algorithm requires the joint processing of all particles' weights, thus preventing a natural concurrency among iterations because the new iterations depend on the previous ones. Second, the communication overhead of particle routing among the processing units (PUs) after resampling can be extensive. This communication, also referred to as *particle routing*, is needed to exchange particles among the PUs after resampling because the PUs that have a surplus of particles need to route the extra particles to the PUs with a shortage of particles. The overhead of communication can result in prohibitive costs in a large parallel computing environment. This is especially true for the wildfire spread application because the size of each particle is large due to the large number of cells that each particle represents.

A basic approach to implementing parallel/distributed particle filtering is the *centralized resampling* approach. In the centralized resampling, two types of nodes are defined, the PU and the central unit (CU). Sampling and weight computation are implemented on PUs since they are independent for different particles. Resampling is performed on a CU due to its sequential nature. The centralized resampling suffers from scalability issues because it relies on a CU.

To support scalable particle-filter-based data assimilation, several distributed resampling schemas have been developed. For example, the distributed resampling with nonproportional allocation (RNA) allows PUs to exchange particles with local neighbors. Specifically, PUs are arranged in a ring topology, and in each iteration, each PU passes a subset of randomly selected particles to its neighbor in the anticlockwise order and then carries out resampling locally. This local resampling schema supports a large degree of parallelism due to data parallelism and the elimination of the centralized resampling step. However, it gives rise to a large number of iterations until full resampling is achieved.

Our previous work studied the routing policies in distributed particle filtering with both the centralized resampling and distributed resampling (Bai *et al.*, 2016). In the centralized resampling, the CU has the full knowledge about the weight distribution of all the particles on different PUs. Based on this global information, we developed two efficient routing policies, named as the *minimal transfer policy* and the *maximal balance policy*. In the distributed resampling schema of RNA, communications are constrained between neighboring PUs. We developed a hybrid particle routing approach that combines the global routing with the local routing to take advantage of both. This approach uses the local routing to ensure scalability and low communication costs and occasionally invokes the global routing to support faster propagation of "good" particles. More details and the performance of the different particle routing methods can be found in the work of Bai *et al.* (2016).

8.8   Sources

The materials in this chapter are derived from the author's previous research. In particular, description of the DEVS-FIRE simulation model (Section 8.3) is derived from Hu *et al.* (2012). The descriptions of the temperature measurement data (Section 8.4) and the data assimilation for dynamic state estimation (Section 8.5) are derived from Xue *et al.* (2012). The spatial partition-based particle filter (Section 8.6) is based on Long and Hu (2017), and the parallel/distributed particle filtering (Section 8.7) is derived from Bai *et al.* (2016).

Chapter 9

Look Ahead

9.1 A New Area of Research and Development

The modeling and simulation field has witnessed research and development in a wide range of topics. Discrete simulations (e.g., discrete event simulation, discrete time simulation, agent-based simulation) have emerged as a major type of simulation and have been applied to a variety of applications. Tremendous progress has been made on both the modeling and simulation fronts, such as the theory of modeling and simulation (Zeigler, 1976; Zeigler *et al.*, 2000) and parallel and distributed simulation (Fujimoto, 2000). These works form the foundation of modeling and simulation to support new research and development.

The rise of data brings new opportunities and challenges to the modeling and simulation field. In particular, integrating real-time data with simulation model represents a new study area that has been largely overlooked by previous work. This study area shows growing importance due to the increasing availability of real-time data and the rising need for using simulation to support real-time decision-making for complex dynamic systems (as exemplified by the growing interests in *digital twin* technologies in recent years). Integrating real-time data with simulation model holds the promise of enabling simulation-based prediction/analysis. Dynamic data-driven simulation (DDDS) presented in this book offers a systematical way to integrate the two for achieving this goal.

This concluding chapter discusses several topics and open issues related to DDDS, with the goal of fostering future research and development in this area. The discussions in this chapter also serve to review the major concepts that are developed in this book. The discussions are organized according to the following topics:

- simulation model,
- measurement data,
- data assimilation,
- external inputs,
- computation cost,
- DDDS applications.

9.2   Simulation Model

Simulation model plays an essential role in DDDS, including: (1) acting as the state transition model that is needed in data assimilation; and (2) enabling simulation runs to support simulation-based prediction/analysis. In data assimilation, the state transition model (i.e., the simulation model) incorporates the knowledge about how a dynamic system works. This knowledge is combined with the information from measurement data to produce state estimates. In simulation-based prediction/analysis, the simulation model is initialized with the real-time state estimated from data assimilation. It runs in parallel with the real system to provide real-time prediction/analysis of the system's future behavior.

The operational usage of a simulation model in DDDS means that the simulation model needs to be developed at an abstraction level corresponding to the real-time operation of a dynamic system. This does not mean that the simulation model must capture every detail of a dynamic system's operation. Developing such a model may not be feasible due to the complexity it involves. A detailed model also increases the dimension of state that makes it more difficult to carry out data assimilation. A topic of future research is to study how data assimilation can work with simulation models at different abstraction levels and develop guidelines for choosing the right models and the right process noises to be used in data assimilation.

The online model calibration allows a simulation model's parameters to be dynamically calibrated based on real-time measurement data. This capability, however, does not mean that one can start from an arbitrary model and expect data assimilation to make it correct. Online model calibration needs to start from a right baseline model. If the parameters of the baseline model are largely wrong or if the structure of the baseline model has major deficiencies, online calibration will not work well. In general, although DDDS works with imperfect simulation models, the quality of model is still important. A poor-quality model not only brings challenges to data assimilation but also directly influences the simulation-based prediction/analysis result. Developing high-quality simulation models remains an essential task.

9.3 Measurement Data

Measurement data are real-time data reflecting a system's dynamic state. In sequential Bayesian filtering, these data are used to compute the likelihood probability for updating the prior distribution of state estimates predicted by the state transition model. The result of the update is the posterior distribution of the state estimates. The mapping from the state space to the measurement data is defined by the measurement model, which can be thought of as a model of sensors used for data collection.

This book focuses on a special type of measurement data called *scalar Gaussian measurements*, which are scalar measurement data (i.e., containing numerical values) with additive Gaussian noises (i.e., the measurement noises follow Gaussian distributions). The likelihood computation for scalar Gaussian measurement is described in Section 6.6. Besides scalar Gaussian measurement, many other types of measurement data exist. For example, an important type of measurement data in discrete event systems is *event-based measurement*, i.e., data indicating the occurrence of an event (e.g., a job is finished). These event-based measurements carry information about the state of a dynamic system. However, more research is needed to assimilate these data in discrete simulations. Another type of measurement data are the data based on human observations and judgments. These data can be referred to as *soft data* to differentiate them from *hard data* that are collected from sensors (Long and Hu, 2014). Soft data carry useful information; however, they may be subjective, biased, or incomplete and thus pose unique challenges in data assimilation.

For complex simulation applications, measurement data may come from different sources and show different characteristics. This brings the need of *data fusion* to integrate data from heterogeneous sources to produce consistent, accurate, and meaningful information. How to integrate data fusion with data assimilation to support DDDS is a future research topic.

The quality of measurement data has a direct impact on data assimilation. The measurement data quality can include multiple aspects, such as measurement noise (i.e., how accurate the measurement data are), frequency of data collection (i.e., how often data are collected), and number of data points in each step (i.e., how many sensors are used). It is also related to the amount of "useful" information the measurement data carry. Depending on where sensors are deployed, the information collected from the different sensors can have different levels of usefulness from the state estimation point of view. For example, wildfire spread is a heterogeneous process due to the non-uniform fuel, terrain, and weather that result in different rates of spread in different regions of a fire area. From the data assimilation point of view, the measurement data collected from the most active regions of a fire would carry more useful information for estimating the state of a wildfire. Given limited sensor resources, a research topic is to study how to deploy the sensors in a way so that they collect the most useful data for data assimilation.

9.4 Data Assimilation

Data assimilation supports dynamic state estimation and online model calibration by providing a systematic way of assimilating real-time data into simulation models. This book covers sequential data assimilation based on the sequential Bayesian filtering framework, with a special focus on data assimilation for discrete simulations. Due to the discrete/hybrid states and non-linear non-Gaussian behavior of discrete simulations, particle filters are used to support data assimilation for discrete simulations. Practical issues of applying particle filter-based data assimilation to discrete simulations are discussed in Chapter 7.

Complex simulation applications often have a large number of state variables. The many state variables result in high dimensional states, which makes it difficult for the data assimilation to converge to the true state. One way to address this issue is to enhance the data assimilation methods. For example, more advanced sampling/ resampling techniques in particle filtering may be developed to support more accurate and robust state estimation. The *spatial partition-based particle filter* described in Section 8.6 is an example of this. Another way to address this issue is to employ state dimension reduction techniques. As an example, the data assimilation for the wildfire spread simulation in Chapter 8 focuses only on the *unburned/burning/burned* state of forest cells while ignoring other state variables. This type of dimension reduction can be useful; nevertheless, more research is needed in order to apply them in systematic ways.

Data assimilation has been studied in other science fields, such as meteorology and oceanography. Although these fields mainly deal with differential equation models, one should pay attention to the data assimilation advances in these fields as they may be adapted and applied to discrete simulations too.

Finally, assimilating data into discrete simulations is still a relatively new topic. The result of particle filter-based data assimilation is influenced by a range of parameters, including the quality of the simulation model, the quality of measurement data (e.g., accuracy, frequency, and number of data points), and data assimilation configurations, such as step length and the number of particles. An important research topic is to systematically study and quantify the impact of these parameters on data assimilation results (see, for example, Gu and Hu, 2010, Huang *et al.*, 2022, and Cho *et al.*, 2020).

9.5   External Inputs

External inputs capture the influence of a dynamic system from its external environment or other systems. External inputs have significant impact on a dynamic system's behavior and thus are an important factor to consider during data assimilation and simulation-based prediction/analysis. This book does not provide in-depth coverage of external input modeling and forecasting. Nevertheless, this topic deserves more attention from the modeling and simulation community in future research.

External inputs may be modeled using probability distributions or simulation modeling that explicitly models the mechanisms of input generation. Alternatively, they may be forecasted based on time series analysis or machine learning approaches. This is an area where data modeling techniques (such as machine learning) can work together with simulation modeling to support more accurate simulation-based prediction/analysis.

9.6   Computation Cost

A unique feature of DDDS is that it works in a real-time context. The real-time setting poses real-time computation requirement for all the activities involved in DDDS. Specifically, since data assimilation is carried out in a stepwise fashion, each step's DDDS activities need to be finished before the next step starts. For applications that have short data assimilation steps and complex simulation models, efficient computation of data assimilation and simulation is essential.

Particle-filter-based data assimilation for discrete simulations typically has a high computation cost due to the large number of particles it uses. Each particle includes a sampling step that involves a simulation. An important way of addressing the high computation costs is to use parallel and distributed computing. Several approaches to parallel/distributed particle filtering are discussed in Section 8.7. However, more research is needed for applying them to discrete simulations. For example, a potential topic is studying how to make parallel/distributed particle filtering work with simulation models that themselves are too complex to run on a single computing node. This calls for research on high-performance computing that integrates parallel/distributed particle filtering with parallel/distributed simulation.

9.7   DDDS Applications

To demonstrate the value of DDDS, it is imperative to apply DDDS to a wide range of applications. The different DDDS applications can not only showcase how DDDS works but also inspire new ideas and suggest new research directions for further development. Several applications, such as the wildfire spread simulation application and the road traffic simulation application, are used throughout the book to illustrate the DDDS concepts. Nevertheless, the application of DDDS is not limited by these examples. Applying DDDS to more applications is one of the major tasks for future development.

Many researchers/practitioners already use real-time data in their simulation applications in one way or another. DDDS provides a framework to reexamine how real-time data may be used in these applications. For example, instead of setting the initial state directly from measurement data, which can result in unrealistic or erroneous states for a simulation model, as discussed in Section 5.4, the dynamic state estimation activity of DDDS provides a way to estimate the initial state using data assimilation. Applying DDDS to these applications can lead to more systematic ways of using real-time data and hence more accurate and robust simulation results.

This book focuses on using DDDS to support real-time prediction/analysis. DDDS may also support other use cases. For example, data assimilation may be used to estimate whether or not a real system is in an undesirable state and thus support online performance monitoring and fault diagnosis for a real system. DDDS can also be a key player to support *digital twin* technologies that have received much attention in recent years. We expect to see more use cases of DDDS as more applications are developed.

References

Arulampalam, S., Maskell, S., Gordon, N., and Clapp, T. (2002). A tutorial on particle filters for on-line non-linear/non-Gaussian Bayesian tracking. *IEEE Transactions on Signal Processing*, 50(2): 174–188.

Bai, F., Gu, F., Hu, X., and Guo, S. (2016). Particle routing in distributed particle filters for large-scale spatial temporal systems. *IEEE Transactions on Parallel and Distributed Systems (TPDS)*, 27(2): 481–493.

Balci, O. (1997). Verification, validation and accreditation of simulation models. *Proceedings of the 29th Conference on Winter Simulation*, pp. 135–141.

Balci, O. (2012). A life cycle for modeling and simulation. *Simulation*, 88: 870–883.

Biller, B., and Gunes, C. (2010). Introduction to simulation input modeling. *Proceedings of 2010 Winter Simulation Conference*. Innsbruck, Austria, February 15–17, pp. 49–58.

Bolic, M., Djuric, P. M., and Hong, S. (2005). Resampling algorithms and architectures for distributed particle filters. *IEEE Transaction on Signal Processing*, 53(7): 2442–2450.

Bonabeau, E. (2002). Agent-based modeling: Methods and techniques for simulating human systems. *Proceedings of the National Academy of Sciences of the USA (PNAS)*, 99(Suppl. 3): 7280–7287.

Briers, M., Doucet, A., and Maskell, S. (2010). Smoothing algorithms for state-space models. *Annals of the Institute Statistical Mathematics*, 62: 61–89.

Brownlee, J. (2017). 4 strategies for multi-step time series forecasting. https://machinelearningmastery.com/multi-step-time-series-forecasting/.

Burks, A. W. (1970). *Essays on Cellular Automata*. Urbana, IL: University of Illinois Press.

Casella, G. and Berger, R. L. (1990). *Statistical Inference*. Belmont, CA: Duxbury Press.

Cho, Y., Huang, Y., and Verbraeck, A. (2020). Strategic use of data assimilation for dynamic data-driven simulation. *Proceedings Computational Science — ICCS 2020 — 20th International Conference*, V. V. Krzhizhanovskaya, G. Závodszky, M. H. Lees, P. M. A. Sloot, J. J. Dongarra, S. Brissos, and J. Teixeira (Eds.), vol. 12142 of Lecture Notes in Computer Science, pp. 31–44.

Courtier, P., Thépaut, J. N., and Hollingsworth, A. (1994). A strategy for operational implementation of 4D-Var, using an incremental approach. *Quarterly Journal of the Royal Meteorological Society*, 120: 1367–1387.

Cruz, M. G. and Alexander, M. E. (2013). Uncertainty associated with model predictions of surface and crown fire rates of spread. *Environmental Modelling & Software*, 47: 16–28.

Davar, S. and Mohammadi, A. (2017). Event-based particle filtering with point and set-valued measurements. *Proceedings 25th European Signal Processing Conference*, Kos, Greece, August 2017, pp. 211–215.

Doucet, A. and Johansen, A. M. (2009). A tutorial on particle filtering and smoothing: Fifteen years later. In *The Oxford Handbook of Nonlinear Filtering*. New York: Oxford University Press.

Doucet, A., de Freitas, N., and Gordon, N. J. (2001). An introduction to sequential Monte Carlo methods. In A. Doucet, J. F. G. de Freitas, and N. J. Gordon (Eds.), *Sequential Monte Carlo Methods in Practice*. New York: Springer-Verlag, pp. 3–13.

Evans, J. R. (2016). *Business Analytics, Methods, Models and Decisions*. Pearson Education Inc.

Evensen, G. (1994). Sequential data assimilation with nonlinear quasi-geostrophic model using Monte Carlo methods to forecast error statistics. *Journal of Geophysical Research*, 99(C5): 143–162.

Evensen, G. (2009). *Data Assimilation: The Ensemble Kalman Filter*. Berlin: Springer.

Forrester, J. W. (1972). *World Dynamics*. Cambridge, MA: Wright-Allen Press.

Fried, J. S., and Fried, B. D. (1996). Simulating wildfire containment with realistic tactics. *Forest Science*, 42(3): 267–281.

Fritzson, P. (2004). *Principles of Object-Oriented Modeling and Simulation with Modelica 2.1*. Wiley-IEEE Press.

Fujimoto, R. M. (2000). *Parallel and Distributed Simulation Systems*. Wiley Interscience.

Gardner, M. (1970). The fantastic combinations of John Conway's new solitaire game 'life'. In *Mathematical Games*. Scientific American. Vol. 223, no. 4, pp. 120–123.

Gordon, N. J., Salmond, D. J., and Smith, A. F. M. (1993). Novel approach to nonlinear/non-Gaussian Bayesian state estimation. *IEE Proceedings F — Radar and Signal Processing*, 140(2): 107–113.

GPS.gov. (n.d.). GPS accuracy. https://www.gps.gov/systems/gps/performance/accuracy/.

Gu, F. and Hu, X. (2010). Analysis and quantification of data assimilation based on sequential Monte Carlo methods for wildfire spread simulation. *International Journal of Modeling, Simulation, and Scientific Computing (IJMSSC)*, 1(4): 445–468.

Hol, J. D., Schon, T. B., and Gustafsson F. (2006). On resampling algorithms for particle filters. Proceedings IEEE nonlinear statistical signal processing workshop, 2006, pp. 79–82.

Houtekamer, P. and Mitchell, H. L. (1998). Data assimilation using an ensemble Kalman filter technique. *Monthly Weather Review*, 126(3): 796–811.

Houtekamer, P. L. and Mitchell, H. L. (2001). A sequential ensemble Kalman filter for atmospheric data assimilation. *Monthly Weather Review*, 129: 123–137.

Hu, X. (2011). Dynamic data driven simulation. *SCS Modeling and Simulation Magazine*, 2: 16–22.

Hu, X. (2022). Data assimilation for simulation-based real-time prediction/analysis. *Proceedings Annual Modeling and Simulation Conference* (ANNSIM 2022), July 2022, San Diego, CA, USA.

Hu, X. and Ntaimo, L. (2009). Integrated simulation and optimization for wildfire containment. *The ACM Transactions on Modeling and Computer Simulation (TOMACS)*, 19(4), pp. 1–29.

Hu, X. and Sun, Y. (2007). Agent-based modeling and simulation of wildland fire suppression. *Proceedings of the 2007 Winter Simulation Conference*, December 2007, Washington, DC, USA.

Hu, X. and Wu, P. (2019). A data assimilation framework for discrete event simulations. *ACM Transactions on Modeling and Computer Simulation (TOMACS)*, 29(3): Article No. 17.

Hu, X., Sun, Y., and Ntaimo, L. (2012). DEVS-FIRE: Design and application of formal discrete event wildfire spread and suppression models. *Simulation*, 88(3): 259–279.

Huang, Y., Xie, X., Cho, Y., and Verbraeck, A. (2022). Particle filter-based data assimilation in dynamic data-driven simulation: sensitivity analysis of three critical experimental conditions. Simulation, Advance online publication. https://doi.org/10.1177/00375497221143988.

Kim, B.S., Kang, B.G., Choi, S.H., and Kim, T.G. (2017). Data modeling versus simulation modeling in the big data era: Case study of a greenhouse control system. *Simulation*, 93(7): 579–594.

Lahoz, W., Khattatov, B., and Menard, R. (Eds.). (2010). *Data Assimilation: Making Sense of Observations.* Springer.

Liu, J. and West, M. (2001). Combined parameter and state estimation in simulation-based filtering. In A. Doucet, J. F. G. de Freitas, and N. J. Gordon (Eds.), *Sequential Monte Carlo Methods in Practice.* New York: Springer-Verlag.

Long, Y. and Hu, X. (2014). Dynamic data driven simulation with soft data. *Proceedings 2014 Spring Simulation Multi-Conference (SpringSim'14),* Symposium on Theory of Modeling and Simulation (TMS/DEVS), April 2014, Tampa, FL, USA.

Long, Y. and Hu, X. (2017). Spatial partition-based particle filtering for data assimilation in wildfire spread simulation. *ACM Transactions on Spatial Algorithms and Systems (TSAS),* 3(2): Article No. 5.

Lorenc, A. C. (1986). Analysis methods for numerical weather prediction. *Quarterly Journal of the Royal Meteorological Society,* 112: 1177–1194.

Mazzocchi, F. (2015). Could big data be the end of theory in science? A few remarks on the epistemology of data-driven science. *EMBO Reports,* 16(10): 1250–1255.

McCaw, W. L., Gould, J. S., and Cheney, N. P. (2008). Existing fire behaviour models under-predict the rate of spread of summer fires in open jarrah (Eucalyptus marginata) forest. *Australian Forestry,* 71(1): 16–26.

Ntaimo, L., Hu, X., and Sun, Y. (2008). DEVS-FIRE: Towards an integrated simulation environment for surface wildfire spread and containment. *Simulation,* 84(4): 137–155.

Ören, T. I. (2001). Impact of data on simulation: From early practices to federated and agent-directed simulations. *Proceedings of the EUROSIM 2001,* 4th International Eurosim Congress, June 2001, Delft, the Netherlands, pp. 3–8.

Pyne, S. J., Andrews, P. L., and Laven, R. D. (1996). *Introduction to Wildland Fire,* 2nd edn. New York: John Wiley & Sons.

Qiu, F., and Hu, X. (2013). Spatial activity-based modeling for pedestrian crowd simulation. *Simulation,* 89(4): 451–465.

Reynolds, C. (1987). Flocks, herds and schools: A distributed behavioral model. *SIGGRAPH'87: Proceedings of the 14th Annual Conference on Computer Graphics and Interactive Techniques.* Association for Computing Machinery, July 1987, Anaheim, California, USA, pp. 25–34.

Reynolds, C. (1999). Steering behaviors for autonomous characters. *Proceedings of Game Developers Conference.* Miller Freeman Game Group, San Francisco, CA, pp. 763–782.

Ross, S. (1997). *Introduction to Probability Models,* 6th edn. Academic Press.

Rothermel, R. C. (1972). A mathematical model for predicting fire spread in wildland fuels. Research Paper INT-115. Ogden, UT: U.S. Department of Agriculture, Forest Service, Intermountain Forest and Range Experiment Station. p. 40.

Sijs, J. and Lazar, M. (2012). Event based state estimation with time synchronous updates. *IEEE Transactions on Automatic Control*, 57(10): 2650–2655.

Sullivan, A. L. (2009a). Wildland surface fire spread modelling, 1990–2007. 1: Physical and quasi-physical models. *International Journal of Wildland Fire*, 18(4): 349–368.

Sullivan, A. L. (2009b). Wildland surface fire spread modelling, 1990–2007. 2: Empirical and quasi-empirical models. *International Journal of Wildland Fire*, 18(4): 369–386.

Thrun, S., Burgard, W., and Fox, D. (2005). *Probabilistic Robotics*. Cambridge, MA, USA: MIT Press.

Van Wagner, C. E. (1973). Height of crown scorch in forest fires. *Canadian Journal of Forest Research*, 3(3): 373–378.

Van Wagner, C. E. (1975). Convection temperatures above low intensity forest fires. *Canadian Forest Service Bi-monthly Research Notes*, 31(2): 21.

Wikipedia. (n.d.). Boids. https://en.wikipedia.org/wiki/Boids.

Wolfram, S. (2002). *A New Kind of Science*. Wolfram Media.

Xie, X. and Verbraeck, A. (2019). A particle filter-based data assimilation framework for discrete event simulations. *Simulation*, 95(11): 1027–1053.

Xue, H., Gu, F., and Hu, X. (2012). Data assimilation using sequential Monte Carlo methods in wildfire spread simulation. *The ACM Transactions on Modeling and Computer Simulation (TOMACS)*, 22(4): Article No. 23.

Zeigler, B. P. (1976). *Theory of Modeling and Simulation*. Wiley Interscience.

Zeigler, B. P. and Sarjoughian, H. S. (2003). Introduction to DEVS modeling and simulation with JAVA: Developing component-based simulation models. Technical report. University of Arizona.

Zeigler, B. P., Praehofer, B., and Kim, T. G. (2000). *Theory of Modelling and Simulation: Integrating Discrete Event and Continuous Complex Dynamic Systems*, 2nd edn. Academic Press.

Zhang, K. (2020). Real-time calibration of large-scale traffic simulators: Achieving efficiency through the use of analytical mode. PhD Dissertation, Massachusetts Institute of Technology.

Index